When atoms are travelling
straight down through empty
space by their own weight, at
quite indeterminate times
and places, they swerve ever
so little from their course,
just so much that you would
call it a change of direction.
If it were not for this swerve,
everything would fall down-
wards through the abyss of
space. No collision would take
place and no impact of atom
on atom would be created.
Thus nature would never have
created anything.
— Lucretius

Swerve Editions

Edited by Jonathan Crary, Sanford Kwinter, and Bruce Mau

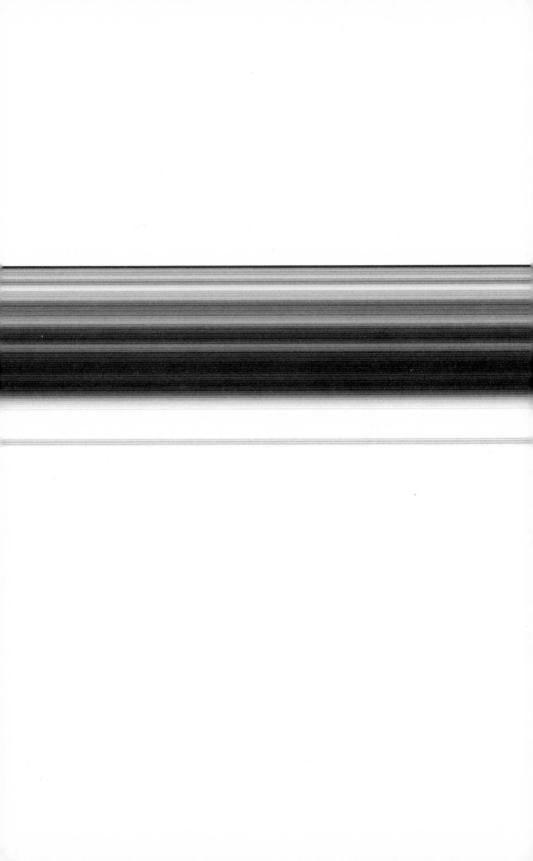

A THOUSAND YEARS OF NONLINEAR HISTORY

Manuel De Landa

Swerve Editions   New York   2000

I would like to dedicate this book to my parents, Manuel De
Landa and Carmen Acosta De Landa. I would also to thank
Celia Schaber for her constant support and inspiration,
Don McMahon for his careful editing and useful suggestions,
and Meighan Gale and the editors at Zone Books.
—Manuel De Landa

First Paperback Edition
Eighth Printing, 2014

Printed in the United States of America.

Distributed by The MIT Press,
Cambridge, Massachusetts, and London, England

Library of Congress Cataloging-in-Publication Data

De Landa, Manuel.
    A Thousand Years of Nonlinear History /
Manuel De Landa.
        p.   cm.
    Includes bibliographical references and index.
    ISBN 978-0-942299-32-9 (paper)
1. Science—Philosophy—History. 2. Nonlinear theories—
History. 3. Philosophy—History. 4. Geology—History.
5. Biology—History. 6. Linguistics—History. I. Title.

Q174.8.D43   1997
501—dc20
                                96-38752

Contents

Introduction

Despite its title, this is not a book of history but a book of philosophy. It is, however, a deeply historical philosophy, which holds as its central thesis that all structures that surround us and form our reality (mountains, animals and plants, human languages, social institutions) are the products of specific historical processes. To be consistent, this type of philosophy must of necessity take real history as its starting point. The problem is, of course, that those who write history, however scholarly, do so from a given philosophical point of view, and this would seem to trap us in a vicious circle. But just as history and philosophy may interact in such a way as to make an objective assessment of reality impossible — when entrenched worldviews and routine procedures for gathering historical evidence constrain each other negatively — they can also interact positively and turn this mutual dependence into a virtuous circle. Moreover, it may be argued that this positive interaction has already begun. Many historians have abandoned their Eurocentrism and now question the very rise of the West (Why not China or Islam? is now a common question), and some have even left behind their anthropocentrism and include a host of nonhuman histories in their accounts. A number of philosophers, for their part, have benefited from the new historical evidence that scholars such as Fernand Braudel and William McNeill have unearthed, and have used it as a point of departure for a new, revived form of materialism, liberated from the dogmas of the past.

Philosophy is not, however, the only discipline that has been influenced by a new awareness of the role of historical processes. Science, too, has acquired a historical consciousness. It is not an exaggeration to say that in the last two or three decades history has infiltrated physics, chemistry, and biology. It is true that nineteenth-century thermodynamics had already introduced time's arrow into physics, and hence the idea of irreversible historical processes. And the theory of evolution had already shown that animals and plants were not embodiments of eternal essences but piecemeal historical constructions, slow accumulations of adaptive traits cemented together via reproductive isolation. However, the classical versions of these two theories incorporated a rather weak notion of history into their conceptual machinery: both classical thermodynamics and Darwinism admitted only one possible historical outcome, the reaching of thermal equilibrium or of the fittest design. In both cases, once this point was reached, historical processes ceased to count. In a sense, optimal design or optimal distribution of energy represented an end of history for these theories.

It should come as no surprise, then, that the current penetration of science by historical concerns has been the result of advances in these two disciplines. Ilya Prigogine revolutionized thermodynamics in the 1960s by showing that the classical results were valid only for closed systems, where the overall quantities of energy are always conserved. If one allows an intense flow of energy in and out of a system (that is, if one pushes it *far from equilibrium*), the number and type of possible historical outcomes greatly increases. Instead of a unique and simple form of stability, we now have multiple coexisting forms of varying complexity (static, periodic, and chaotic *attractors*). Moreover, when a system switches from one stable state to another (at a critical point called a *bifurcation*), minor fluctuations may play a crucial role in deciding the outcome. Thus, when we study a given physical system, we need to know the specific nature of the fluctuations that have been present at each of its bifurcations; in other words, we need to know its history to understand its current dynamical state.[1]

And what is true of physical systems is all the more true of biological ones. Attractors and bifurcations are features of any system in which the dynamics are not only far from equilibrium but also *nonlinear*, that is, in which there are strong mutual interactions (or feedback) between components. Whether the system in question is composed of molecules or of living creatures, it will exhibit endogenously generated stable states, as well as sharp transitions between states, as long as there is feedback and an intense flow of energy coursing through the system. As biology begins to include these nonlinear dynamical phenomena in its models—for example, the mutual stimulation involved in the case of evolutionary "arms races" between predators and prey—the notion of a "fittest design" will lose its meaning. In an arms race there is no optimal solution fixed once and for all, since the criterion of "fitness" itself changes with the dynamics.[2] As the belief in a fixed criterion of optimality disappears from biology, real historical processes come to reassert themselves once more.

Thus, the move away from energetic equilibrium and linear causality has reinjected the natural sciences with historical concerns. This book is an exploration of the possibilities that might be opened to philosophical reflection by a similar move in the social sciences in general and history in particular. These pages explore the possibilities of a nonlinear and nonequilibrium history by tracing the development of the West in three historical narratives, each starting roughly in the year 1000 and culminating in our own time, a thousand years later. But doesn't this approach contradict my stated goal? Isn't the very idea of following *a line of development*, century by century, inherently linear? My answer is that a nonlinear conception of history has absolutely nothing to do with a style of presentation, as if one could truly capture the nonequilibrium dynamics of human historical processes by jumping back and forth among the centuries. On the contrary, what is needed here is not a textual but a physical operation: much as history has infiltrated physics, we must now allow physics to infiltrate human history.

Earlier attempts in this direction, most notably in the pioneering work of the physicist Arthur Iberall, offer a useful illustration of the conceptual shifts that this infiltration would involve. Iberall was perhaps the first to visualize the major transitions in early human history (the transitions from hunter-gatherer to agriculturalist, and from agriculturalist to city dweller) *not* as a linear advance up the ladder of progress but as the crossing of nonlinear critical thresholds (bifurcations). More specifically, much as a given chemical compound (water, for example) may exist in several distinct states (solid, liquid, or gas) and may switch from stable state to stable state at critical points in the intensity of temperature (called *phase transitions*), so a human society may be seen as a "material" capable of undergoing these changes of state as it reaches critical mass in terms of density of settlement, amount of energy consumed, or even intensity of interaction.

Iberall invites us to view early hunter-gatherer bands as gas particles, in the sense that they lived apart from each other and therefore interacted rarely and unsystematically. (Based on the ethnographic evidence that bands typically lived about seventy miles apart and assuming that humans can walk about twenty-five miles a day, he calculates that any two bands were separated by more than a day's distance from one another.[3]) When humans first began to cultivate cereals and the interaction between human beings and plants created sedentary communities, humanity liquefied or condensed into groups whose interactions were now more frequent although still loosely regulated. Finally, when a few of these communities intensified agricultural production to the point where surpluses could be harvested, stored, and redistributed (for the first time allowing a division of labor between producers and consumers of food), humanity acquired a crystal state, in the sense that central governments now imposed a symmetrical grid of laws and regulations on the urban populations.[4]

However oversimplified this picture may be, it contains a significant clue as to the nature of nonlinear history: if the different "stages" of human history were indeed brought about by phase transitions, then they are not "stages" at all—that is, progressive developmental steps, each better than the previous one, and indeed leaving the previous one behind. On the contrary, much as water's solid, liquid, and gas phases may coexist, so each new human phase simply added itself to the other ones, coexisting and interacting with them without leaving them in the past. Moreover, much as a given material may solidify in alternative ways (as ice or snowflake, as crystal or glass), so humanity liquefied and later solidified in different forms. The nomads of the Steppes (Huns, Mongols), for example, domesticated animals not plants, and the consequent pastoral lifestyle imposed on them the need to move with their flocks, almost as if they had condensed not into a pool of liquid but into a moving, at times turbulent, fluid. When these nomads did acquire a solid state (during the reign of Genghis Khan, for instance), the resulting structure was more like glass than crystal, more amorphous and less centralized. In other words, human history did not follow a straight line, as if everything pointed toward civilized societies as humanity's ultimate goal. On the contrary, at each bifurcation alternative stable states were possible, and once actualized, they coexisted and interacted with one another.

I am aware that all we have here are suggestive metaphors. It is the task of the various chapters of this book to attempt to remove that metaphorical content. Moreover, even as metaphors, Iberall's images suffer from another drawback: inorganic matter-energy has a wider range of alternatives for the generation of structure than just these simple phase transitions, and what is true for simple "stuff" must be all the more so for the complex materials that form human cultures. In other words, even the humblest forms of matter and energy have the potential for *self-organization* beyond the relatively simple type involved in the creation of crystals. There are, for instance, those coherent waves called *solitons*, which form in many different types of materials, ranging from ocean waters (where they are called tsunamis) to lasers. Then there are the aforementioned stable states (or attractors), which can sustain coherent cyclic activity of different types (periodic or chaotic).[5] Finally, and unlike the previous examples of nonlinear self-organization where true innovation cannot occur, there is what we may call "nonlinear combinatorics," which explores the different combinations into which entities derived from the previous processes (crystals, coherent pulses, cyclic patterns) may enter. It is from these unlimited combinations that truly novel structures are generated.[6] When put together, all these forms of spontaneous structural generation suggest that inorganic matter is much more variable and creative than we ever imagined. And this insight into matter's inherent creativity needs to be fully incorporated into our new materialist philosophies.

While the concept of self-organization, as applied to purely material and energetic systems, has been sharpened considerably over the last three decades, it still needs to be refined before we can apply it to the case of human societies. Specifically, we need to take into account that any explanation of human behavior must involve reference to irreducible intentional entities such as "beliefs" and "desires," since expectations and preferences are what guide human decision making in a wide range of social activities, such as politics and economics. In some cases the decisions made by individual human beings are highly constrained by their position and role in a hierarchical organization and are, to that extent, geared toward meeting the goals of that organization. In other cases, however, what matters is not the planned results of decision making, but the *unintended collective consequences* of human decisions. The best illustration of a social institution that emerges spontaneously from the interaction of many human decision makers is that of a pre-capitalist market, a collective entity arising from the decentralized interaction of many buyers and sellers, with no central "decider" coordinating the whole process. In some models, the dynamics of markets are governed by periodic attractors, which force markets to undergo boom-and-bust cycles of varying duration, from three-year business cycles to fifty-year-long waves.

Whether applied to self-organized forms of matter-energy or to the unplanned results of human agency, these new concepts demand a new methodology, and it is this methodological change that may prove to be of philosophical significance. Part of what this change involves is fairly obvious: the equations scientists use to model nonlinear processes cannot be solved by hand, but demand the use of computers. More technically, unlike linear equations (the type most prevalent in science), nonlinear ones are very difficult to solve *analytically*, and demand the use of detailed numerical simulations carried out with the help of digital machines. This limitation of analytical tools for the study of nonlinear dynamics becomes even more constraining in the case of nonlinear combinatorics. In this case, certain combinations will display *emergent properties*, that is, properties of the combination as a whole which are more than the sum of its individual parts. These emergent (or "synergistic") properties belong to the *interactions between parts,* so it follows that a top-down analytical approach that begins with the whole and dissects it into its constituent parts (an ecosystem into species, a society into institutions), is bound to miss precisely those properties. In other words, analyzing a whole into parts and then attempting to model it by *adding up* the components will fail to capture any property that emerged from complex interactions, since the effect of the latter may be multiplicative (e.g., mutual enhancement) and not just additive.

Of course, analytical tools cannot simply be dismissed due to this inherent limitation. Rather, a top-down approach to the study of complex entities needs to be *complemented* with a bottom-up approach: analysis needs to go hand in hand with synthesis. And here, just as in the case of nonlinear dynamics, computers offer an indispensable aid. For example, instead of studying a rain forest top down, starting from the forest as a whole and dividing it into species, we unleash within the computer a population of interacting virtual "animals" and "plants" and attempt to generate from their interactions whatever systematic properties we ascribe to the ecosystem as a whole. Only if the resilience, stability, and other properties of the whole (such as the formation of complex food webs) emerge spontaneously in the course of the simulation can we assert that we have captured the nonlinear dynamics and combinatorics of rain forest formation. (This is, basically, the approach taken by the new discipline of Artificial Life.[7])

In this book, I attempt a philosophical approach to history which is as bottom-up as possible. This does not mean, of course, that every one of my statements has emerged after careful synthetic simulations of social reality. I do take into account the results of many bottom-up simulations (in urban and economic dynamics), but research in this direction is still in its infancy. My account is bottom-up in that I make an effort not to postulate systematicity when I cannot show that a particular system-generating process has actually occurred. (In particular, I refrain from speaking of society as a whole forming a system and focus instead on subsets of society.) Also, I approach entities at any given level (the level of nation-states, cities, institutions, or individual decision makers) in terms of *populations* of entities at the level immediately below.

Methodologically, this implies a rejection of the philosophical foundations of orthodox economics as well as orthodox sociology. Although the former (neoclassical microeconomics) begins its analysis at the bottom of society, at the level of the individual decision maker, it does so in a way that atomizes these components, each one of which is modeled as maximizing his or her individual satisfaction ("marginal utility") in isolation from the others. Each decision maker is further atomized by the assumption that the decisions in question are made on a case-by-case basis, constrained only by budgetary limitations, ignoring social norms and values that constrain individual action in a variety of ways. Orthodox sociology (whether functionalist or Marxist-structuralist), on the other hand, takes society as a whole as its point of departure and only rarely attempts to explain in detail the exact historical processes through which collective social institutions have emerged out of the interactions among individuals.

Fortunately, the last few decades have witnessed the birth and growth of a synthesis of economic and sociological ideas (under the banner of "neoinstitutional economics"), as exemplified by the work of such authors as Douglas North, Viktor Vanberg, and Oliver Williamson. This new school (or set of schools) rejects the atomism of neoclassical economists as well as the holism of structuralist-functionalist sociologists. It preserves "methodological individualism" (appropriate to any bottom-up perspective) but rejects the idea that individuals make decisions solely according to self-interested (maximizing) calculations, and instead models individuals as rule followers subjected to different types of normative and institutional constraints that apply collectively. Neoinstitutionalism rejects the "methodological holism" of sociology but preserves what we may call its "ontological holism," that is, the idea that even though collective institutions emerge out of the interactions among individuals, once they have formed they take on "a life of their own" (i.e., they are not just reified entities) and affect individual action in many different ways.[8]

Neoinstitutionalist economists have also introduced sociological concepts into economics by replacing the notion of "exchange of goods" with the more complex one of "transaction," which brings into play different kinds of collective entities, such as institutional norms, contracts, and enforcement procedures. Indeed, the notion of "transaction" may be said to add to linear economics some of the "friction" that its traditional models usually leave out: imperfections in markets due to limited rationality, imperfect information, delays and bottlenecks, opportunism, high-cost enforceability of contracts, and so on. Adding "transaction costs" to the classical model is a way of acknowledging the continuous presence of non-linearities in the operation of real markets. One of the aims of the present book is to attempt a synthesis between these new ideas and methodologies in economics and the corresponding concepts in the sciences of self-organization.[9]

In Chapter One I approach this synthesis through an exploration of the history of urban economics since the Middle Ages. I take as my point of departure a view shared by several materialist historians (principally Braudel and McNeill): the specific dynamics of European towns were one important reason why China and Islam, despite their early economic and technological lead, were eventually subjected to Western domination. Given that an important aim of this book is to approach history in a non-teleological way, the eventual conquest of the millennium by the West will not be viewed as the result of "progress" occurring there while failing to take place outside of Europe, but as the result of certain dynamics (such as the mutually stimulating dynamics involved in arms races) that intensify the accumulation of knowledge and technologies, and of certain institutional norms and organizations. Several different forms of mutual stimulation (or of "positive feedback," to use the technical term) will be analyzed, each involving a different set of individuals and institutions and evolving in a different area of the European urban landscape. Furthermore, it will be argued that the Industrial Revolution can be viewed in terms of reciprocal stimulation between technologies and institutions, whereby the elements involved managed to form a closed loop, so that the entire assemblage became self-sustaining. I refer to this historical narrative as "geological" because it concerns itself exclusively with dynamical elements (energy flow, nonlinear causality) that we have in common with rocks and mountains and other nonliving historical structures.

Chapter Two addresses another sphere of reality, the world of germs, plants, and animals and hence views cities as ecosystems, albeit extremely simplified ones. This chapter goes beyond questions of inanimate energy flow, to consider the flows of organic materials that have informed urban life since the Middle Ages. In particular, it considers the flow of food, which keeps cities alive and in most cases comes from outside the town itself. Cities appear as parasitic entities, deriving their sustenance from nearby rural regions or, via colonialism and conquest, from other lands. This chapter also considers the flow of genetic materials through generations — not so much the flow of human genes as those belonging to the animal and plant species that we have managed to domesticate, as well as those that have constantly eluded our control, such as weeds and microorganisms. Colonial enterprises appear in this chapter not only as a means to redirect food toward the motherland, but also as the means by which the genes of many nonhuman species have invaded and conquered alien ecosystems.

Finally, Chapter Three deals with the other type of "materials" that enter into the human mixture: linguistic materials. Like minerals, inanimate energy, food, and genes, the sounds, words, and syntactical constructions that make up language accumulated within the walls of medieval (and modern) towns and were transformed by urban dynamics. Some of these linguistic materials (learned, written Latin, for example) were so rigid and unchanging that they simply accumulated as a dead structure. But other forms of language (vulgar, spoken Latin) were dynamic entities capable of giving birth to new structures, such as French, Spanish, Italian, and Portuguese. This chapter traces the history of these emergences, most of them in urban environments, as well as of the eventual rigidification (through standardization) of the dialects belonging to regional and national capitals, and of the effects that several generations of media (the printing press, mass media, computer networks) have had on their evolution.

Each chapter begins its narrative in the year 1000 A.D. and continues (more or less linearly) to the year 2000. Yet, as I said above, despite their style of presentation, these three narratives do not constitute a "real" history of their subjects but rather a sustained philosophical meditation on some of the historical processes that have affected these three types of "materials" (energetic, genetic, and linguistic). The very fact that each chapter concentrates on a single "material" (viewing human history, as it were, from the point of view of that particular material) will make these narratives hardly recognizable as historical accounts. Yet, most of the generalizations to be found here have been made by historians and are not the product of pure philosophical speculation.

In the nonlinear spirit of this book, these three worlds (geological, biological, and linguistic) will *not* be viewed as the progressively more sophisticated stages of an evolution that culminates in humanity as its crowning achievement. It is true that a small subset of geological materials (carbon, hydrogen, oxygen, and nine other elements) formed the substratum needed for living creatures to emerge and that a small subset of organic materials (certain neurons in the brain) provided the substratum for language. But far from advancing in stages of increased perfection, these successive emergences were—and will be treated here as—mere accumulations of different types of materials, accumulations in which each successive layer does not form a new world closed in on itself but, on the contrary, results in coexistences and interactions of different kinds. Besides, each accumulated layer is animated from within by self-organizing processes, and the forces and constraints behind this spontaneous generation of order are common to all three.

In a very real sense, reality is *a single matter-energy* undergoing phase transitions of various kinds, with each new layer of accumulated "stuff" simply enriching the reservoir of nonlinear dynamics and nonlinear combinatorics available for the generation of novel structures and processes. Rocks and winds, germs and words, are all different manifestations of this dynamic material reality, or, in other words, they all represent the different ways in which this single matter-energy *expresses itself*. Thus, what follows will not be a chronicle of "man" and "his" historical achievements, but a philosophical meditation on the history of matter-energy in its different forms and of the multiple coexistences and interactions of these forms. Geological, organic, and linguistic materials will all be allowed to "have their say" in the form that this book takes, and the resulting chorus of material voices will, I hope, give us a fresh perspective on the events and processes that have shaped the history of this millennium.

1: LAVAS A

ND MAGMAS

Geological History 1000–1700 A.D.

We live in a world populated by structures—a complex mixture of geological, biological, social, and linguistic constructions that are nothing but accumulations of materials shaped and hardened by history. Immersed as we are in this mixture, we cannot help but interact in a variety of ways with the other historical constructions that surround us, and in these interactions we generate novel combinations, some of which possess emergent properties. In turn, these synergistic combinations, whether of human origin or not, become the raw material for further mixtures. This is how the population of structures inhabiting our planet has acquired its rich variety, as the entry of novel materials into the mix triggers wild proliferations of new forms.

In the organic world, for instance, soft tissue (gels and aerosols, muscle and nerve) reigned supreme until 500 million years ago. At that point, some of the conglomerations of fleshy matter-energy that made up life underwent a sudden *mineralization*, and a new material for constructing living creatures emerged: bone. It is almost as if the mineral world that had served as a substratum for the emergence of biological creatures was reasserting itself, confirming that geology, far from having been left behind as a primitive stage of the earth's evolution, fully coexisted with the soft, gelatinous newcomers. Primitive bone, a stiff, calcified central rod that would later become the vertebral column, made new forms of movement control possible among animals, freeing them from many constraints and literally setting them into motion to conquer every available niche in the air, in water, and on land. And yet, while bone allowed the complexification of the animal phylum to which we, as vertebrates, belong, it never forgot its mineral origins: it is the living material that most easily petrifies, that most readily crosses the threshold back into the world of rocks. For that reason, much of the geological record is written with fossil bone.

The human endoskeleton was one of the many products of that ancient mineralization. Yet that is not the only geological infiltration that the human species has undergone. About eight thousand years ago, human populations began mineralizing again when they developed an urban *exoskeleton*: bricks of sun-dried clay became the building materials for their homes, which in turn surrounded and were surrounded by stone monuments and defensive walls. This exoskeleton served a purpose similar to its internal counterpart: to control the movement of human flesh in and out of a town's walls. The urban exoskeleton also regulated the motion of many other things: luxury objects, news, and food, for example. In particular, the weekly markets that have always existed at the heart of most cities and towns constituted veritable motors, periodically concentrating people and goods from near and faraway regions and then setting them into motion again, along a variety of trade circuits.[1]

Thus, the urban infrastructure may be said to perform, for tightly packed populations of humans, the same function of motion control that our bones do in relation to our fleshy parts. And, in both cases, adding minerals to the mix resulted in a fantastic combinatorial explosion, greatly increasing the variety of animal and cultural designs. We must be careful when drawing these analogies, however. In particular, we must avoid the error of comparing cities to organisms, especially when the metaphor is meant to imply (as it has in the past) that both exist in a state of internal equilibrium, or homeostasis. Rather, urban centers and living creatures must be seen as *different* dynamical systems operating far from equilibrium, that is, traversed by more or less intense flows of matter-energy that provoke their unique metamorphoses.[2]

Indeed, urban morphogenesis has depended, from its ancient beginnings in the Fertile Crescent, on intensification of the consumption of nonhuman energy. The anthropologist Richard Newbold Adams, who sees social evolution as just another form that the self-organization of energy may take, has pointed out that the first such intensification was the cultivation of cereals.[3] Since plants, via photosynthesis, simply convert solar energy into sugars, cultivation increased the amount of solar energy that traversed human societies. When food production was further intensified, humanity crossed the bifurcation that gave rise to urban structures. The elites that ruled those early cities in turn made other intensifications possible—by developing large irrigation systems, for example—and urban centers mutated into their imperial form. It is important to emphasize, however, that cereal cultivation was only one of several possible ways of intensifying energy flow. As several anthropologists have pointed out, the emergence of cities may have followed *alternative routes to intensification*, as when the emergence of urban life in Peru fed off a reservoir of fish.[4] What matters is not agriculture per se, but the great increase in the flow of matter-energy through society, as well as the transformations in urban form that this intense flow makes possible.

From this point of view cities arise from the flow of matter-energy, but once a town's mineral infrastructure has emerged, it reacts to those flows, creating a new *set of constraints* that either intensifies or inhibits them. Needless to say, the walls, monumental buildings, streets, and houses of a town would make a rather weak set of constraints if they operated on their own. Of course, they do not. Our historical exploration of urban dynamics must therefore include an analysis of the *institutions* that inhabit cities, whether the bureaucracies that run them or the markets that animate them. Although these institutions are the product of collective human decision making, once in place they also react back on their human components to limit them and control them, or, on the contrary, to set them in motion or accelerate their mutation. (Hence institutions constitute a set of emergent positive and negative constraints, but on a smaller scale.)

The birth of Europe, around the eleventh century of our era, was made possible by a great agricultural intensification. As Lynn White, Jr., a historian of medieval technology, has shown, in the centuries preceding the second millennium, "a series of innovations occurred which consolidated to form a remarkably efficient new way of exploiting the soil."[5] These innovations (the heavy plow, new ways of harnessing the horse's muscular energy, the open-field system, and triennial field rotation) were mutually enhancing as well as interdependent, so that only when they fully meshed were their intensifying effects felt. The large increase in the flow of energy created by this web of technologies allowed for the reconstitution of the European exoskeleton, the urban framework that had for the most part collapsed with the Roman Empire. Beginning around 1000 A.D., large populations of walled towns and fortified castles appeared in two great zones: in the south, along the Mediterranean coast, and in the north, along the coastlands lying between the trade waters of the North Sea and the Baltic.

As city historians often point out, urbanization has always been a discontinuous phenomenon. Bursts of rapid growth are followed by long periods of stagnation.[6] The wave of accelerated city building that occurred in Europe between the eleventh and thirteenth centuries is no exception. Many of the great towns in the north, such as Brussels and Antwerp, were born in this period, and the far older cities of Italy and the Rhineland experienced enormous growth. This acceleration in urban development, however, would not be matched for another five hundred years, when a new intensification in the flow of energy—this time arising from the exploitation of fossil fuels—propelled another great spurt of city birth and growth in the 1800s. Interestingly, more than the proliferation of factory towns made possible by coal, the "tidal wave of medieval urbanization"[7] laid out the most enduring features of the European urban structure, features that would continue to influence the course of history well into the twentieth century.

There are two basic processes by which cities can emerge and grow. A town may develop spontaneously, acquiring its irregular shape by following the topographical features of the landscape, or it may inherit its shape from the distribution of villages that have amalgamated to form it. Such was the case of medieval Venice, which accounts for its labyrinthine streets. On the other hand, a city may be the result of conscious planning; a regular, symmetrical form may be imposed on its development, to facilitate orderly settlement. During the deceleration that followed the year 1300, the relatively few new cities that were born were of the latter type, perhaps reflecting the increasing political centralization of the time. Versailles, with its grid of broad avenues converging at the center of power, is a perfect illustration. However, the difference between self-organized and planned cities is not primarily one of form, but of the decision-making processes behind the genesis and subsequent development of that form. That is, the crucial distinction is between centralized and decentralized decision making in urban development. There are towns that have been purposefully designed to mimic the "organic" form of curvilinear streets, and there are towns whose grid-patterned streets evolved spontaneously, due to some peculiarity of the environment. Furthermore, most cities are mixtures of the two processes:

> If we were to scan several hundred city plans at random across the range of history, we would discover a more fundamental reason to question the usefulness of urban dichotomies based on geometry. We would find that the two primary versions of urban arrangement, the planned and the "organic", often exist side by side.... In Europe, new additions to the dense medieval cores of historic towns were always regular.... Most historic towns, and virtually all those of metropolitan size, are puzzles of premeditated and spontaneous segments, variously interlocked or juxtaposed.... We can go beyond. The two kinds of urban form do not always stand in contiguous relationship. They metamorphose. The reworking of prior geometries over time leaves urban palimpsests where a once regular grid plan is feebly ensconced within a maze of cul-de-sacs and narrow winding streets.[8]

The mineralization of humanity took forms that were the combined result of conscious manipulation of urban space by some central agency and of the activities of many individuals, without any central "decider." And yet, the two processes, and the forms they typically give rise to, remain distinct despite their coexistence and mutual transformations. On the one hand, the grid is "the best and quickest way to organize a homogeneous population with a single social purpose."[9] On the other hand, whenever a *heterogeneous* group of people comes together spontaneously, they tend to organize themselves in an interlocking urban pattern that interconnects them without *homogenizing* them.

Even though from a strictly physical viewpoint accelerations in city building are the result of intensifications in the flow of energy, the actual form that a given town takes is determined by human decision making. A similar distinction between centralized and decentralized decision making must be made with respect to the social institutions that determine how energy flows through a city—that is, with respect to the city's "distribution systems."[10] On the one hand, there are bureaucracies, hierarchical structures with conscious goals and overt control mechanisms. On the other, there are peasant and small-town markets, self-organized structures that arise spontaneously out of the activities of many individuals, whose interests only partially overlap. (I have in mind here a place in a town where people gather every week, as opposed to markets in the modern sense: dispersed collections of consumers served by many middlemen.)[11]

Bureaucracies have always arisen to effect a planned extraction of energy surpluses (taxes, tribute, rents, forced labor), and they expand in proportion to their ability to control and process those energy flows. Markets, in contrast, are born wherever a regular assembly of independent decision makers gathers, whether at church or at the border between two regions, presenting individuals with an opportunity to buy, sell, and barter. The distinction between these two types of energy distribution systems exactly parallels the one above, only on a smaller scale. One system sorts out human beings into the internally homogeneous ranks of a bureaucracy. The other brings a heterogeneous collection of humans together in a market, where their complementary economic needs enmesh.

Markets and bureaucracies are, however, more than just collective mechanisms for the allocation of material and energetic resources. When people exchanged goods in a medieval market, not only resources changed hands but also *rights of ownership*, that is, the rights to use a given resource and to enjoy the benefits that may be derived from it.[12] Hence, market transactions involved the presence of collective institutional norms (such as codes of conduct and enforceable contracts). Similarly, medieval bureaucracies were not only organizations that controlled and redistributed resources via centralized commands, they themselves were sets of mutually stabilizing institutional norms, a nexus of contracts and routines constituting an apparatus for collective action. The rules behind bureaucracies tended to be more formalized than the informal

conventions and codes of conduct behind markets, and more importantly, they tended to become a "constitution," that is, a set of contracts
defining a homogeneous, common enterprise not easily disaggregated
into a set of heterogeneous bilateral contracts like those involved in market transactions.[13]

Markets and bureaucracies, as well as unplanned and planned cities,
are concrete instances of a more general distinction: self-organized *meshworks* of diverse elements, versus *hierarchies* of uniform elements. But
again, meshworks and hierarchies not only coexist and intermingle, they
constantly give rise to one another. For instance, as markets grow in size
they tend to form commercial hierarchies. In medieval times this was
true of the great fairs, such as the Champagne fairs of the thirteenth century, which came to have as many participants as most towns had inhabitants: "If a fair is envisaged as a pyramid, the base consists of the many
minor transactions in local goods, usually perishable and cheap, then
one moves up to the luxury goods, expensive and transported from far
away. At the very top of the pyramid came the active money market without which business could not be done at all—or at any rate not at the
same pace."[14]

Thus, once markets grew past the size of local, weekly gatherings, they
were ranked and organized from the top, giving rise to a hybrid form:
a hierarchy of meshworks. The opposite hybrid, a meshwork of hierarchies, may be illustrated by the system of power in the Middle Ages. Urban
bureaucracies were but one of a number of centralized institutions that
coexisted in the Middle Ages. Royal courts, landed aristocracies, and
ecclesiastical hierarchies all entered into complex, uneasy mixtures. There
was never a "super-elite" capable of globally regulating the mix, so local
constraints (shifting alliances, truces, legal debates) worked alongside
formal procedures in generating stability. If we add to this the fact that
the state and the church in the West arose from heterogeneous origins
(unlike China or Islam where all these hierarchical structures had emerged
within a homogeneous cultural tradition), the system of power in the early
part of this millennium was a true mesh of hierarchical organizations.[15]

Meshworks and hierarchies need to be viewed not only as capable of
giving rise to these complex hybrids but also as in constant interaction with
one another. Primitive bureaucracies had evolved in the Middle Ages to
regulate certain aspects of market life (for instance, to arbitrate disputes
between markets when their catchment areas overlapped), or to provide
security for the big fairs. However, we must not imagine that the mere
existence of a command hierarchy meant that the global rules of a
bureaucracy could in practice be enforced. In medieval times, the norms

that governed economic life—the norms that guaranteed that contracts would be honored or that measures, weights, and currencies would remain stable—were for the most part not global, but based on self-defense, retaliation, and other local controls. As one historian has put it, the enforcement of economic norms in the Middle Ages was a combination of centralized decision making and a "self-regulating mechanism compounded by a balance of terror and a lively sense of mutual advantage felt by all members of the international community."[16]

The large populations of towns and cities that emerged in Europe after the year 1000 may be classified by their relative proportions of meshwork and hierarchical components. By far the majority of settlements were small towns, with more market than command ingredients in their mix. Over half of all European urban dwellers lived in those local market centers, even though each town had fewer than two thousand residents. Then came intermediate-sized towns (fewer than ten thousand inhabitants), which began adding local and regional administrative functions and, hence, a higher proportion of command components. Control of roads and supervision of travelers, two centralized functions absent from small towns, were already practiced here. A wider variety of institutional forms inhabited those larger settlements: courts, jails, hospitals, religious foundations. But as complexity increased, so did rarity: while there were about 3,000 small towns in northern Europe, there were only 220 of intermediate size.[17] Denser urban concentrations were even rarer, but for the same reason sustained a wider range of functions:

> Cities with more than 10,000 residents stood out in Medieval Europe, except in northern Italy and Flanders where the spread of cloth production and the increase in trade permitted relatively intense urbanization. Elsewhere, large size was correlated with complex administrative, religious, educational, and economic functions. Many of the big towns—for example, Barcelona, Cologne or Prague—supported universities as well as a wide variety of religious institutions. Their economies were diversified and included a wide range of artisans and service workers.... The large cities of 1330 owed their size to the multiplicity of their functions.... The same point can be made about the few urban giants of the Middle Ages. Paris, Milan, Venice, and Florence were commercial and manufacturing cities, and also political capitals.[18]

This multiplicity of urban centers, internally differentiated by size and complexity, can be compared to other populations of towns that emerged elsewhere. Urbanization explosions had occurred in Islam and China at least two centuries before those in Europe. But in those two regions, cities and towns had to compete with a larger sociopolitical entity that emerged only later in the West: the central state. While Islam in the early part of the millennium had some towns (Córdova, Ceuta) similar to those in the West, huge towns, such as Baghdad or Cairo, that housed royal hierarchies were the rule there.[19] China, too, showed a greater percentage of towns subjected to a central authority than autonomous towns defined by the movement of people and goods through their walls. William McNeill is one of several historians who think that one of the reasons for the West's eventual domination of the millennium lies in the different mixtures of centralized and decentralized decision making in its towns:

> The fact that China remained united politically from Sung to modern times ... is evidence of the increased power government personnel wielded. Discrepancies between the ideas of the marketplace and those of government were real enough; but as long as officials could bring overriding police power whenever they were locally or privately defied, the command element in the mix remained securely dominant.... For this reason the autocatalytic character that European commercial and industrial expansion exhibited between the eleventh and the nineteenth century never got started in China.[20]

In short, McNeill's hypothesis is that explosive, self-stimulating ("autocatalytic") urban dynamics cannot emerge when hierarchical components overwhelm meshwork components. Fernand Braudel seems to agree with this hypothesis when he asserts the existence of a "dynamic pattern of turbulent urban evolution in the West, while the pattern of life in cities in the rest of the world runs in a long, straight and unbroken line across time."[21] One example of the nonlinear, runaway nature of autocatalytic dynamics in many medieval Western towns is the sequence of intensifications of energy flow that propelled urban growth. First came an agricultural intensification causing massive increases in population and therefore giving birth to many cities. Then, as in ancient times, the interaction of these urban centers further intensified energy consumption. One of these intensifications was achieved by harnessing the energy of running water to power grain mills and trip-hammers in forges and to facilitate the fulling of cloth. This was, without exaggeration, an eleventh-century industrial revolution, fueled by solar (agricultural) and gravitational (water) energy.[22]

In addition to raw energy, the turbulent dynamics to which both McNeill and Braudel refer were associated with the intensification of another flow: the flow of money. Howard Odum, a systems ecologist, has developed a theory of money that, though perhaps too simple, offers a useful image here. Money, Odum says, is like energy, only it runs in the opposite direction: energy flows from agricultural villages to the towns they feed, while money flows from town to countryside, to pay for the food. "The flow of energy makes possible the circulation of money [including the energy spent on paperwork, banking, closing deals] and the manipulation of money can control the flow of energy."[23] To apply Odum's schema to medieval life we need to bring our mixtures of market and command ingredients to bear. Contrary to what may be supposed, monetary systems are of not commercial but political origin. Specifically, they were developed by central hierarchies to facilitate the extraction of agricultural surpluses and the raising of taxes.[24] In the early part of the millennium, feudal landlords extracted this excess energy, and in many cases peasants would come to a market town to sell their goods, not to buy other goods, but to get cash to pay their rent to the owners of their land.[25] With that qualification, Odum's idea is useful: monetary flows regulate (inhibit or intensify) energy flows, particularly when the flow of money escapes total control by the state.

Money is best defined as a catalyst or stimulant of trade (and its absence, an inhibitor). Barter, the exchange of goods for goods, is relatively inefficient in that people must wait for their complementary needs to meet. The occasions when one person has exactly the good that the other needs, and vice versa, are exceedingly rare. But any good that is highly desirable and can easily be put back into circulation can play the role of money: blocks of salt, cowry shells, coral, ivory—even cigarettes in modern prisons.[26] Any one of a number of widely desired goods can spontaneously become money simply by being able to flow faster and more easily. And once such self-organized money comes into existence, complementary demands can be meshed together at a distance, greatly increasing the intensity of market exchanges. Frequently coexisting with this spontaneous money are *monetary systems*, with their hierarchy of homogeneous metal coins of different denominations, a system that is not self-organized but planned and implemented by an elite. Planned money, since its inception in ancient Egypt, has used metals as its physical vehicle because they can be weighed and measured, uniformly cut, and standardized.[27]

Whenever these two types of money—the planned and the spontaneous—came into contact, standardized money would inevitably win, causing devaluation of the other, increases in its reserves, and catastrophic inflation. This situation would arise time and again over the centuries, particularly when Europe began colonizing the world. However, in the first few centuries of the millennium the situation was reversed: early Europe was, in a manner of speaking, a colony of Islam, an empire that not only had a more advanced monetary system, but also had invented many of the instruments of credit (from bills of exchange to promissory notes and checks). As Braudel says, "If Europe finally perfected its money, it was because it had to overthrow the domination of the Muslim world."[28] Venice, Florence, Genoa, and other large medieval cities started coining their own copper, silver, and gold money, and the volume of European trade began to rise. From then on, this new flow, catalyzing and control- ling the flow of energy, never ceased accelerating the pace of European history. The flow of money could itself be intensified, either by increas- ing the exploitation of mines, and hence the reservoir of metal, or by speeding up its circulation. These two intensifications, of the volume and velocity of money, affected each other, since "as precious metals became more plentiful coins passed more quickly from hand to hand."[29]

These intense flows of energy and monetary catalysts fueled the great urban acceleration in medieval Europe and kept the towns that made up Europe's great exoskeleton in a turbulent dynamical state. Although large accumulations of money created new commercial hierarchies, the net result was a decrease in the power of central states and a concomitant increase in the autonomy of cities. The intensity of the flows themselves, and not any special feature of the "European psyche" (calculating ratio- nality, say, or a spirit of thrift), is what kept the mixture of market and command components in the right proportions to foster autocatalytic dynamics.[30] One more element must be added to this explanation, how- ever, but this will involve going beyond a conception of markets (and bureaucracies) as allocation mechanisms for scarce resources.

This point might be clarified by applying certain ideas recently devel- oped by the neoinstitutionalist economist Douglas North. As we noted above, not only resources change hands in the marketplace but also property rights; hence the market facilitates simple exchanges as well as potentially complex transactions. The latter involves a host of "hidden" costs ranging from the energy and skill needed to ascertain the quality of a product, to the drawing of sales and employment contracts, to the enforcement of those contracts. In small medieval markets these "trans- action costs" were minimal, and so were their enforcement mechanisms: threats of mutual retaliation, ostracism, codes of conduct, and other informal constraints sufficed to allow for the more or less smooth func- tioning of a market. But as the volume and scale of trade intensified (or as its character changed, as in the case of foreign, long-distance trade), new institutional norms and organizations were needed to regulate the flow of resources, ranging from standardized weights and measures to the use of notarial records as evidence in merchant law courts or state courts. North's main point is that, as medieval markets grew and complexified, their transaction costs increased accordingly; without a set of institutional norms and organizations to keep those costs down, the turbulent intensification of trade in the West would have come to a halt. Economies of scale in trade and low-cost enforceability of contracts were, according to North, mutually stimulating.[31]

Many institutional norms emerged in an unplanned way—those related to common law or to informal codes of conduct, for example—and slowly "sedimented" within towns in the Middle Ages. Others, such as printed lists of prices or maritime insurance schemes, were deliberately introduced to reduce transaction costs by improving the flow of market information or by spreading the risks of large investments. Those cities engaging in types of trade with particularly high transaction costs, such as long-distance trade, seem to have been the incubators of many institutional innovations. As these "cultural materials" (informal constraints, formal rules, enforcement procedures) acting as trade catalysts accumulated, they began to diffuse through the urban environment. As North observes, "Merchants carried with them in long-distance trade codes of conduct, so that Pisan laws passed into the sea codes of Marseilles. Oleron and Lübeck gave laws to the north of Europe, Barcelona to the south of Europe, and from Italy came the legal principle of insurance and bills of exchange."[32]

One difference between the neoinstitutionalist approach and the one I am trying to sketch here is this: beyond the level of the individual organization, the neoinstitutionalist does not seem to envision yet another emergent larger-scale entity but simply refers to "society" or "the polity" as a whole. This, however, runs the risk of introducing too much homogeneity into our models and of suggesting that human societies form a "totality," that is, an entity on a higher ontological plane than individual institutions and individual human beings. By contrast, speaking of concrete cities (instead of "society" in the abstract) enables us to include in our models historically emergent wholes that do not form totalities but simply larger-scale individual entities. It also reduces the danger of taking too much social uniformity for granted. Individual cities (and nation states) are easier to visualize as encompassing a variety of communities within their borders, and if, as a matter of empirical fact, a given city (or nation-state) displays a high degree of cultural homogeneity, this itself becomes something to be modeled as the result of concrete historical processes. We have already seen that, depending on the mixture of centralized and decentralized decision making behind a city's birth and growth, we can expect different degrees of uniformity and diversity in its infrastructural layout. To this it must be added that, depending on the role that a city plays in the larger urban context in which it functions, the "cultural materials" that accumulate within it will exhibit different degrees of homogeneity and heterogeneity. Specifically, a city may play the role of political capital for a given region and encourage a certain degree of uniformity in its own culture and in that of the smaller towns under its command. On the contrary, a city may act as a *gateway to foreign cultures*, promoting the entry and diffusion of heterogeneous materials that increase its diversity and that of the cities in close contact with it. In either case, viewing cities as individuals allows us to study the interactions between them and the emergent wholes that may result from those interactions.

That groups of cities may form hierarchical structures is a well-known fact at least since the 1930s, when the term "Central Place" system was introduced to refer to pyramids of urban centers. More recently, urban historians Paul Hohenberg and Lynn Hollen Lees have suggested that in addition to hierarchical structures, cities in Europe also formed a meshwork-like assemblage, which they refer to as the "Network System." Let's examine some of the defining traits of these two types of city assemblages, beginning with the Central Place system, exemplified in the Middle Ages by the hierarchies of towns that formed under strong regional capitals such as Paris, Prague, and Milan. As we saw before, the population of towns in medieval Europe was divided by the size and complexity of its individual units. This distribution of sizes was not accidental but directly related to the links and connections between settlements. Much as small towns offered the surrounding countryside a variety of commercial, administrative, and religious services, the towns themselves looked to the more diversified larger cities for services that were unavailable locally. This created pyramids of towns organized around hierarchical levels of complexity. The distribution in space of these hierarchical systems was directly tied to geographical distance, since the residents of a town would only travel so far in search of a desired service. A number of such pyramidal structures arose in the Middle Ages, each organizing a broad, more or less clearly defined region. Generally, the flows of traded goods that circulated up and down these hierarchies consisted of basic necessities, such as food and manufactured products.

In contradistinction, the circulation of *luxury items* originated *somewhere else*. Long-distance trade, which has since Antiquity dealt with prestige goods, is the province of cities outside the Central Place system, cities that act as gateways to faraway trading circuits, as well as nodes in a network not directly constrained by distance. For example, many European gateway cities were maritime ports, connected (more than separated) by the Mediterranean and the Baltic and North Seas.[33] These urban centers formed, according to Hohenberg and Lees, a Network system:

> The Network System, with quite different properties, complements the Central Place System. Instead of a hierarchical nesting of similar centers, distinguished mainly by the number and rarity of services offered, it presents an ordering of functionally complementary cities and urban settlements. The key systemic property of a city is nodality rather than centrality.... Since network cities easily exercise control at a distance, the influence of a town has little to do with propinquity and even less with formal command over territory. The spatial features of the Network System are largely invisible on a conventional map: trade routes, junctions, gateways, outposts.[34]

Instead of a hierarchy of towns, long-distance trading centers formed a meshwork, an interlocking system of complementary economic functions. This is not to imply, however, that all the nodes in the meshwork were of equal importance. Certain economic functions (especially those giving rise to innovations) formed a privileged core within a given network, while others (e.g., routine production tasks) characterized its peripheral zones. Yet, the core of the Network system differed from the acme of the Central Place pyramid. In particular, the influence of a network's main city was more precarious than that of the Central Place, whose dominance tended to be stable. Core cities tended to replace one another in this role, as the intensity of exchange in a given trade route varied over time, or as erstwhile luxury goods (pepper, sugar) became everyday necessities: "Since [these] cities are links in a network, often neither the source nor the ultimate destination of goods, they are in some measure interchangeable as are the routes themselves."[35] Roughly, the sequence of cores was (from the fourteenth to the twentieth centuries) Venice, Antwerp, Genoa, Amsterdam, London, New York.[36] The two systems coexisted, with Central Place towns usually belonging to the middle zone (or semiperiphery) of the Network system.[37]

One very important feature of Central Place and Network systems is the type of *cultural structures* they give rise to. As with many other structures, the raw materials (in this case, cultural habits and norms) need to accumulate slowly and then consolidate, as more or less permanent links are established among them. Hierarchical constructions tend to undergo a homogenization before their materials harden into a pyramid, while meshworks articulate heterogeneous elements, interlocking them without imposing uniformity:

> On one level, the Central Place System serves a homogeneous people well settled in its historical lands. The national capital distills and formalizes the common folk culture and reinjects the civilized product back into local life.... [This contrasts] with the rootless cosmopolitanism of the Network System, with its sharp cultural discontinuities between city and country and between core and periphery.... Core values and techniques are superimposed on a traditional periphery with no attempt at integration or gradual synthesis.[38]

Even before the advent of national capitals, the dominant cities of Central Place hierarchies performed their homogenizations at the regional level, transforming local cultures into "great traditions," as they engaged in book printing and publishing as well as schooling. Gateway cities, on the other hand, helped diffuse heterogeneous elements from alien cultures, as when medieval Venice introduced into Europe products, technology, and architecture from the East. Later on, the cities of the Network system would propagate the ideas of humanism, enlightenment, and radical thought, while giving refuge to persecuted thinkers and publishing forbidden books.[39] The circulation and processing of "cultural materials" through these two different systems of cities are as important in the long run as the mind-sets of the inhabitants of the towns themselves. The latter are, of course, an active element in the mix, to the extent that psychological structures, once they have come into being, affect the dynamics of decision making and hence the flows of energy and money, knowledge and ideas. But what is crucial to emphasize here is that the entire process does not emanate from some essence housed within people's heads, particularly not any reified essence such as "rationality."

In the original version of Central Place theory, created by Walter Christaller in the early 1930s, the human capacity for making maximally efficient decisions (what is now called "optimizing rationality") was taken for granted. The model of Christaller also assumed a frictionless world, where geography lacked irregularities, wealth and power were distributed evenly, and the levels of demand for city services, as well as the distances people would be willing to travel to get them, remained fixed. In this linear world, particular spatial distributions of cities of different rank resulted, as the different centers arranged themselves to minimize travel time for a given service, thus optimizing their collective benefit, or utility.[40] In nonlinear dynamical models of city development, such as those created by Peter Allen and Dimitrios Dendrinos, urban patterns do not result from some global optimizer (such as superrational human decision makers minimizing transportation costs) but from a dynamics of cooperation and conflict among cities, involving growth and decay of centers. In these models, urban settlements grow by attracting population from surrounding rural areas, with job availability and income acting as incentives to immigration while congestion and pollution act as disincentives. Although in principle several cities could share these human resources more or less evenly, the models show a strong tendency for some urban centers to grow at the expense of others and for large centers to inhibit the growth of similarly scaled towns in their vicinity. Moreover, the emergence of stable patterns of coexisting centers seems related to a decrease in the strength and number of direct interactions among towns: too much connectivity (as when every city in the model interacts with every other one) leads to unstable patterns, while decreased connectivity within a hierarchy of towns (that is, fewer interactions between ranks than within a given rank) leads to stability.[41]

Contemporary studies in nonlinear urban dynamics teach us that, in many cases, friction (delays, bottlenecks, conflict, uneven distribution of resources) plays a crucial role in generating self-organization. Hence, eliminating it from our models (by postulating an optimizing rationality, for instance) automatically eliminates the possibility of capturing any real dynamical effect. This insight is even more important when we consider the dynamics of the institutions that channel the flow of energy through cities: markets and bureaucracies. The classical picture of the market, Adam Smith's "invisible hand" model, is just like Christaller's model of urban patterns. It operates in a world completely devoid of friction, where monopolies do not exist and agents are endowed with perfect foresight and have access to costless and unlimited information. Smith's model (or more exactly, its implementation in neoclassical economics) also generates patterns that maximize the benefits to society as a whole, that is, patterns in which supply and demand interact so as to reach optimal equilibrium, precluding wasteful excesses or deficits. This type of market dynamics is, of course, a fiction. And yet this picture of a "rational" free-market dynamics emanating from the interaction of selfish agents reaching optimal conclusions about alternative uses of scarce resources is still at the core of modern linear economics.

Nonlinear approaches to market dynamics, in contrast, emphasize the role of uncertainty in decision making and the inherent costs of information gathering. Imperfect knowledge, incomplete assessment of feedback, limited memory and recall, as well as poor problem-solving skills result in a form of rationality that attains not optimal decisions but more or less satisfactory compromises between conflicting constraints.[42] This "satisficing" or "bounded" rationality, proceeds in many cases by rules of thumb and other adaptive behavioral patterns. This does not preclude some coherence among an agent's expectations, needs, and actions, but it does call for a dynamic explanation of the formation of adequate beliefs, as opposed to simply assuming static forms of rationality. Moreover, it emphasizes that the responses of economic agents in the marketplace are not uniform, that some agents will act more coherently than others, and that the adequacy of their decisions will vary from time to time.[43]

A nonlinear model of market dynamics differs greatly from Adam Smith's. In particular, instead of a single, static equilibrium toward which markets are supposed to gravitate, the nonlinear model allows for multiple dynamical forms of stability. For example, markets may get caught in cyclical equilibriums that force them to undergo successive periods of growth and decay. Hence markets may be both self-regulating and non-optimal.[44] These issues are all the more important when considering medieval markets, which had to cope not only with the effects of imperfect foresight, but with a multiplicity of other nonlinearities: agrarian hierarchies exacting a portion of production, taking it out of circulation; craftsmen selling their products speculatively; money supply affecting prices; and so on. Nonetheless, by the twelfth century, prices throughout Europe fluctuated in unison, and this is what above all characterizes a self-regulating market economy.[45] This collective oscillation, this massive rhythmical breathing across the cities that made up the Central Place and Network systems, can now be captured through the use of nonlinear models, where the impediments created by bounded rationality play a constructive role.[46]

One may think that the suboptimal compromises to which medieval markets were condemned derived from the decentralized nature of their decision-making processes. But a similar conclusion may be reached vis-à-vis centralized bureaucracies, even though their formalized plans and well-defined goals would seem to be products of an optimizing rationality. But here, too, decision making takes place in a world full of uncertainties. Any actual system of information processing, planning, and control will never be optimal but merely practical, applying rote responses to recurrent problems and employing a variety of contingency tactics to deal with unforeseen events. Some of the flows of matter and energy in and out of cities — flows that medieval hierarchies were supposed to regulate — received more attention while others were overlooked and mismanaged.

For instance, by the thirteenth century London had already generated a specialized bureaucracy for handling the flow of water into the city; but management of the flow of waste out of the city did not come about until the nineteenth century, even though the English capital had had recurrent sewage crises since the 1370s. It was not until the river Thames's capacity to transport waste reached its limits, causing an odor that made parliamentary sessions impossible to conduct, that the problem was confronted. Before that, the approach to sewage management had been reactive, unplanned, and piecemeal—hardly optimal.[47]

Thus, to understand the role of decision making in the creation of social order, we need to concentrate not so much on the more or less rational character of *individual* decisions, but on the dynamics (centralized or decentralized) among many interacting decision makers. The hierarchies and meshworks that develop from these interactions (particular bureaucracies, individual markets) in turn become elements of other homogeneous and heterogeneous structures (capitals or gateways), which in turn go on to form Central Place and Network systems. At each level, different nonlinear dynamics take place, with their own multiple equilibriums and bifurcations between alternative stable states. Hence, individual decision making, while important, is simply one element in the mix, interacting and influencing dynamics on only one of a number of scales.[48]

But even at the individual level, what matters is not any particular psychological structure (rationality) so much as problem-solving skills, rules of thumb, and routine procedures, that is, "cultural materials" that can *accumulate over time* within a town's walls. Indeed, many preindustrial cities may be seen as large reservoirs of skills and routines. Those cities recruited from the countryside artisans possessing the most varied abilities and trades, and they were constantly struggling to steal this valuable "human capital" away from each other. To maintain and increase their reservoirs, towns attracted a flow of craftsmen, as well as a variety of professionals, who brought with them skills and procedures that could now be taught to others or imitated, and hence added to the existing stock. As these cultural materials accumulated, they mixed in various ways, forming novel meshworks and hierarchies.

On one hand, the ruling elites of many towns created, between the twelfth and the fifteenth centuries, the guild system, through which they organized all craft activity within the city. Each guild brought together the skills that formed a given trade, and homogenized the means of their transmission by regulating training methods and certification procedures. As skills accumulated and began interacting with one another, trades began to diversify and multiply: "In Nuremberg ... the metal working guilds ... had divided, as early as the thirteenth century, into several dozen independent professions and trades. The same process occurred in Ghent, Strasbourg, Frankfurt and Florence, where the woolen industry, as elsewhere, became a collection of trades. In fact it would be true to say that the boom of the thirteenth century arose out of this newly created division of labor as it proliferated."[49] On the other hand, as specialties multiplied so did the interactions between individual trades, and this gave rise to meshworks of small producers, "symbiotic collections of little enterprises," as the urbanist Jane Jacobs has called them.[50]

While the big gateway cities at the core of the Network system, as well as those at the top of Central Place pyramids, gave rise to elaborate hierarchies of guilds and ever more rigid regulations, towns inhabiting the middle zone (that is, not too small to be condemned to remain a supply region for the core), engaged in what Jacobs calls "import-substitution dynamics." Instead of simply exchanging raw materials for manufactured goods from the big cities, the artisans of these towns developed the skills necessary to slowly replace those imports with local production. These new, less regulated skills, in turn, began forming meshworks, as they interlocked with one another in functional complementarity.[51]

The market dynamics of these middle-zone towns were self-stimulating because the money saved by replacing some imports could be spent on new imports, which in turn generated a new round of substitutions. As Jacobs puts it, these small medieval towns, and their small producers, "were forever producing new exports for one another—bells, dyes, buckles, parchment, lace, needles, painted cabinet work, ceramics, brushes, cutlery, paper, sieves and needles, sweetmeats, elixirs, files, pitchforks, sextants—replacing them with local production, becoming customers for still more innovations."[52] Jacobs describes the autocatalytic dynamics that produced these humble goods as evolving through bifurcations, as a critical mass of potentially replaceable imports accumulated within a town, giving rise to a new explosive episode of import replacement. The innovations that came out of this process did not have to be glamorous or highly visible; what mattered was the generation of new skills and the consequent complexification of the meshwork.

Computer simulations of economic meshwork dynamics have shown that, at a certain critical level of complexity, a kind of "industrial takeoff" occurs in the interlocked system of functions constituting the meshwork.[53] Jacobs has gathered evidence indicating that this is indeed the way in which the economy of Europe took off at the turn of the first millennium. At the time, Constantinople was at the top of the urban hierarchy, and Venice (which by the fourteenth century was the metropolis at the core of the Network system) was one of its humble supply zones. The Venetians sold timber and salt to the capital, in exchange for manufactured products. In the eleventh century, however, the economy of Venice began to grow explosively, as a meshwork of small producers began substituting locally manufactured goods for those previously imported from Constantinople. Since the local goods were necessarily rough and primitive by the standards of the capital, Venice could only trade its new surplus products with other backward cities. (Thus, this type of autocatalysis involves not single cities but teams of cities.) In this way, the economy of Venice took off and propelled the city to a position as dominant center. Because the smaller towns that now imported Venetian products were also reservoirs of flexible skills, they eventually created their own import-substitution meshworks. Such was the case of Antwerp, which began as a Venetian supply region for wool; by the fifteenth century it too had become a core of the Network. London had to wait until the nineteenth century before becoming the Network core, but since the Middle Ages it had been substituting imported leather goods from Córdova, to sell to other backward cities.[54]

This kind of volatile trade among small towns should be added to our list of autocatalytic processes animating medieval Europe. Large towns, on the other hand, gave rise to a different type of turbulent dynamics, based on luxury goods (instead of everyday items) involving big firms (instead of small producers), and on strategies that did not rely on the existence of heterogeneous skills. As Braudel says, the proliferation of new trades, and the resultant microspecializations, always characterized the bottom layers of the trade hierarchy. Big business in the Middle Ages, and for centuries afterward, had its own dynamics, which ran in the exact opposite direction: "Even a shopkeeper who made his fortune, and became a merchant, immediately moved out of specialization into non-specialization... obeying the rules of trade at its upper levels. To become and above all to remain a wholesaler meant having not only the right but the duty to handle, if not everything, at any rate as much as possible."[55]

The advantage that nonspecialization gave to these early capitalists was *freedom of motion*, which allowed them to handle any flow of goods that became highly profitable, and to move in and out of flows as their profitability changed. This freedom of choice has characterized capitalism throughout the millennium. The merchants and financiers (and later industrialists) who inhabited the upper levels of the trade hierarchy never invaded low-profit zones. With the exclusion of cash crops for the luxury market, food production and processing were left untouched until the seventeenth century. The same is true of transportation, until the railroads, and of the construction industry, until our century (if we exclude factories and public buildings). If we add to this the retailing of goods, we may conclude that none of the flows of energy and matter that are indispensable for an urban center were penetrated by large commercial hierarchies (and their centralized decision making) until relatively recently.

Even in this age of huge multinational corporations, the command element in the commercial mixture is far from 100 percent. The economist John Kenneth Galbraith, who sharply differentiates between spontaneous economic activity (markets) and planned economic processes (big business), calculates that today roughly half of the Western economy has been taken over by capitalist hierarchies. The other half comprises the low-profit regions, which these hierarchies willingly abandon to the market. According to Galbraith, what gives capitalism this freedom of motion is economy of scale, which is why since the Middle Ages commercial capitalism has been associated with wholesale and not retail. A large firm is better able to absorb shocks and fluctuations and create the plans and strategies that may win it a degree of independence from market forces, indeed the ability *to control and manipulate* those forces to a certain degree.

Such considerations led Braudel to the startling conclusion that "we should not be too quick to assume that capitalism embraces the whole of western society, that it accounts for every stitch in the social fabric... that our societies are organized from top to bottom in a 'capitalist system.' On the contrary... there is a dialectic still very much alive between capitalism on one hand, and its antithesis, the 'non-capitalism' of the lower level on the other."[56] And he adds that, indeed, capitalism was carried upward and onward on the shoulders of small shops and "the enormous creative powers of the market, of the lower story of exchange.... [This] lowest level, not being paralysed by the size of its plant or organization, is the one readiest to adapt; it is the seedbed of inspiration, improvisation and even innovation, although its most brilliant discoveries sooner or later fall into the hands of the holders of capital. It was not the capitalists who brought about the first cotton revolution; all the new ideas came from enterprising small businesses."[57]

There is a misconception, widely shared by economists and philosophers on either side of the political spectrum, that capitalism developed in several stages, being at first competitive and subservient to market forces and only later, in the twentieth century, becoming monopolistic. However, starting in the thirteenth century, capitalists engaged in various noncompetitive practices, in order to create the large accumulations of money that have always characterized the upper levels of the trade pyramid. As we discussed, the early medieval fairs, the meeting points of rich merchants from all over Europe, were veritable hierarchies of meshworks, in which the luxury and money markets dominated the upper echelons. Neither in the long-distance trade of prestige goods nor in the worlds of precious metals and credit did supply and demand reign supreme. On the contrary, most fortunes in these areas were made by the manipulation of these market forces through a variety of noncompetitive practices. There was, of course, intense competition among rich merchants and families, much as today large corporations compete with one another, but these rivalries among oligopolies are fundamentally different from the kind of "anonymous competition" in which small producers and traders engage.[58]

From the Middle Ages to the nineteenth century, not only did individual businesses engage in monopolistic practices, entire cities did too, even groups of cities. By means of noncompetitive practices, a town could greatly aid its merchants and financiers, protecting them from foreign rivals, and stimulating the accumulation of money within its walls. The medieval cities that controlled the Mediterranean and the Baltic and North Seas financed much of their growth from manipulation of markets and by acquiring exclusive control of certain flows, such as spices and silks from the Levant in the case of Venice, or salt in the case of Lübeck. With a monopoly on luxury goods, won and maintained by military force, fourteenth-century Venice dominated the cities around it, not only the small towns constituting its supply regions but other giant towns, such as Florence and Milan. In the north, between the thirteenth and fifteenth centuries, cities like Lübeck and Bruges formed a meshwork of cities known as the Hanseatic League, which was capable of collective action without a centralized organization behind it. The league also engaged in monopolistic practices to trap the towns within its zone of economic influence in a web of supervision and dependence.[59]

We will return shortly to other forms of market manipulation which, according to Braudel, have always characterized certain commercial institutions since the Middle Ages. This will make clear how wrong it is to assume (as many economists to the right and center of the political spectrum tend to do) that market power is something that may be dismissed or that needs to be studied only in relation to some aberrant institutional forms such as overt monopolies. But certain conceptions from the left (particularly the Marxist left) also need to be corrected, in particular, a teleological conception of economic history in terms of a *linear progression* of modes of production. In this Braudel explicitly agrees with Gilles Deleuze and Félix Guattari: capitalism could have arisen anywhere and long before it did in Europe.[60] Its emergence must be pictured as a bifurcation, a phase transition that might have taken place somewhere else had the conditions been right (for instance, in the huge camel caravans along the Silk Road in the thirteenth century).[61] Moreover, the institutions that emerged after this bifurcation must be viewed not as replacing previous institutions (i.e., markets) but as fully coexisting with them without forming a societywide "system." It is true that prices across Europe were pulsating to the same rhythm from medieval times and this gave the entire continent a certain economic coherence (sometimes referred to as a "world-economy"), but it would be a mistake to confuse world-economies with the "capitalist system," since India, China, and Islam also formed coherent economic areas (as powerful as those of Europe) without giving rise to capitalism.[62]

The conceptual confusion engendered by all the different uses of the word "capitalism" (as "free enterprise" or as "industrial mode of production" or, more recently, as "world-economy") is so entrenched that it makes an objective analysis of economic power almost impossible. One could, of course, simply redefine the term "capitalism" to include "power to manipulate markets" as a constitutive part of its meaning and to rid it of some of its teleological connotations. But as philosophers of science know well, when a theory begins redefining its terms in an ad hoc way to fit the latest round of negative evidence, it shows by this very act that it has reached the limits of its usefulness. In view of this, it would seem that the only solution is to replace this tired word with a neologism, perhaps the one Braudel suggested, "antimarkets," and to use it exclusively to refer to a certain segment of the population of commercial and industrial institutions.[63]

In addition to monopolies, the most obvious form of manipulation of supply and demand, preindustrial antimarkets used several other mechanisms to further their accumulations and increase their domination. For example, goods bought directly from a producer at a low price were often stored in large warehouses until the market price rose to a desired level. Market prices sometimes increased of their own accord, as happened during wars, but whenever they did not the merchants who owned these huge reservoirs could artificially inflate prices, perhaps by buying certain amounts of a given product at a high price (or, vice versa, deflate prices by dumping lower-priced goods).[64] Long-distance trade was another means to free oneself of the laws and limitations of the local market. In terms of volume, long-distance luxury trade was minuscule in comparison to the flows of humble goods that circulated in the medieval markets. But what it lacked in one form of intensity it made up in another:

Long-distance trade certainly made super-profits: it was after all based on the price differences between two markets very far apart, with supply and demand in complete ignorance of each other and brought into contact only by the activities of the middleman. There could only have been a competitive market if there had been plenty of separate and independent middlemen. If, in the fullness of time competition did appear, if super-profits vanished from one line, it was always possible to find them again on another route with different commodities. If pepper became commonplace and declined in value, tea, coffee, or calicoes were waiting in the wings to take the place of the former prima donna.[65]

Such was the freedom of movement that characterized antimarkets, a freedom made possible by extensive *credit*. Much as primitive or metallic money was a catalyst for small-scale commercial exchange, credit was the great accelerator for antimarket transactions, both wholesale and long-distance trade. Credit represented one more form of the autocatalytic or turbulent dynamics that propelled preindustrial European cities ahead of their Eastern rivals, eventually enabling Europe to dominate the rest of the world. Credit (or, more exactly, compound interest) is an example of explosive, self-stimulating growth: money begetting money, a diabolical image that made many civilizations forbid usury. European merchants got around this prohibition through the use of the "bill of exchange," originally a means of long-distance payment (inherited from Islam); as it circulated from fair to fair its rate of return accrued usuriously. (This disguised form of usury was tolerated by church hierarchies due to the many risks the circulation of bills of exchange involved.) The flow of credit — and the institutions that grew around this flow, such as banks and stock exchanges — was crucial for self-sustained economic growth at the top, and it was one more flow antimarket institutions monopolized early on.[66]

To return to European urban history, the deceleration of urban expansion that followed the year 1300 had a variety of effects. The birthrate of new towns decreased significantly, as did continuous growth across the full spectrum of city sizes. In the subsequent four centuries many small towns disappeared, and only the larger towns continued to grow. In a sense, the long depression acted as a selection pressure, favoring the large and hence increasing the proportion of command elements in the mix. Simultaneously, the first nation-states began to consolidate, in regions previously organized by Central Place hierarchies, as the dominant cities, some of which became *national capitals*, began to swallow up and discipline the towns in their orbit. The gateway cities that made up the Network system lost some of their autonomy yet continued to grow, becoming *maritime metropolises*. Hence, while relatively few towns were born in this period, the existing population of cities changed significantly. The capital and the metropolis, and the huge concentrations of people they housed, became increasingly visible features of the European urban structure.

Anne Querrien has described the characteristics typical of these two types of large towns, while warning us that in reality a pure capital or metropolis is rare, that more often than not we are dealing with mixtures. A metropolis, she says, is like "a membrane which allows communication between two or more milieus, while the capital serves as a nucleus around which these milieus are rigorously organized."[67] Metropolitan centers exercise their influence across international boundaries, while capitals are the guardians and protectors of these frontiers and the territories they encompass. Hence, while the former arise by the sea, the latter are often landlocked, bound to their hinterland. Capitals tend to place restrictions on the flows of trade and use taxes, tolls, and tariffs to extract energy from these circuits; conversely, metropolitan cities tend to free these fluxes of all obstacles, seeking to exploit their distant peripheries more thoroughly. (We have here two different forms of power, xenophobic nationalism versus salt-water imperialism.)[68] In the period of nation-state formation, Paris, Madrid, Baghdad, and Peking were perfect examples of national capitals, while Venice, Genoa, Córdova, and Canton typified the maritime metropolis. Cities such as London were mixtures of both types.

The emergence of powerful nation-states, and the concomitant decrease in autonomy of the cities they absorbed (and even of the city-states that remained independent), could have brought the different forms of self-stimulating dynamics we have described to a halt. That this did not happen was due to yet one more form of autocatalysis unique to the West: continued *arms races*. The historian Paul Kennedy has argued that this type of self-stimulation depended in turn on the fact that the nations of Europe, unlike China or Islam, were never able to form a single, homogeneous empire, and have remained until today a meshwork of hierarchies. It was within this meshwork that advances in offensive weaponry stimulated innovations in defense technology, leading to an ever-growing armament spiral:

> While this armament spiral could already be seen in the manufacture of crossbows and armor plate in the early fifteenth century, the principle spread to experimentation with gunpowder weapons in the following fifty years. It is important to recall here that when cannon were first employed, there was little difference between the West and Asia in their design and

effectiveness.... Yet, it seems to have been only in Europe that the impetus existed for constant improvement: in the gunpowder grains, in casting
much smaller (yet equally powerful) cannon from bronze and tin alloys, in
the shape and texture of the barrel and the missile, in the gun mountings
and carriages.[69]

These arms races had a variety of consequences. They affected the
mineralization of Europe, as the new mobile siege artillery made the simple high walls that surrounded most towns obsolete. Fortification changed
radically, as town walls were built lower while becoming more elaborate,
now incorporated into complex assemblages of ditches, ramparts, para-
pets, and covered passageways. This had important consequences for
the cities enclosed within these fortified walls. Before 1520, when a town
outgrew its mineral membrane, the wall could be easily disassembled and
reconstructed farther away. But now, the new star-shaped systems of
defense that had replaced it were prohibitively expensive to move, so that
the towns so fortified were thereafter condemned to grow vertically.[70] On
the other hand, the new fortress designs, as well as the artillery that had
catalyzed them into existence, began to consume a rapidly increasing
share of a town's wealth. This favored nations over city-states, since only
the former could sustain the intensification of resource exploitation that
the new technologies demanded.

Kennedy has added his voice to the chorus of historians who, having
rejected Eurocentrism, came to realize that even as late as 1500 China or
Islam was much better positioned to dominate the millennium than Europe. (Hence, the fact that Europe managed to do this against the odds
warrants explanation.) Many of the inventions that Europeans used to colonize the world (the compass, gunpowder, paper money, the printing
press) were of Chinese origin, while Europe's accounting techniques and
instruments of credit (which are often cited as examples of her unique
"rationality") came from Islam. Thus, nothing intrinsic to Europe determined the outcome, but rather a dynamics bearing no inherent relationship to any one culture. In this, Kennedy agrees with Braudel and
McNeill: an excess of centralized decision making in the East kept turbulent dynamics under control, while they raged unobstructed in the West.
To be sure, at several points in her history Europe could have become a
unified hierarchy, and this would have ground these dynamics to a halt.
This happened in the sixteenth century with the Hapsburg Empire, and
later on with the rise of Napoleon and Hitler. Yet all these efforts proved
abortive, and European nations remained a meshwork.

Perhaps the most damaging effect of centralization was that it made

Eastern nations too dependent on the individual skills of their elites. Sometimes these skills were in short supply, as in the Ottoman Empire after 1566, when it was ruled by thirteen incompetent sultans in succession. Because of the excess of command element in the mix, as Kennedy says, "an idiot sultan could paralyze the Ottoman Empire in a way that a pope or Holy Roman emperor could never do for all of Europe."[71] In a similar way, China's outlook was turned inward by its elite at a crucial point in history, when the secret to world domination lay in the conquest of the oceans, both for the profits of long-distance trade and for the flows of energy and materials that colonization made possible.

China had an early lead in the naval race, having successfully pioneered expeditions to the Indian Ocean as early as 1405, in which her "largest vessels probably displaced about 1,500 tons compared to the 300 tons of Vasco da Gama's flagship...at the end of the same century. Everything about these expeditions eclipsed the scale of later Portuguese endeavors. More ships, more guns, more manpower, more cargo capacity...."[72] However, China's rigid elite turned back its outward-looking policies and turned the country inward. Had China's expeditions continued, "Chinese navigators might well have rounded Africa and discovered Europe before Prince Henry the Navigator died."[73] And European cities might have found themselves colonies and supply regions of a faraway empire.

Those were the dangers and missed opportunities that too much centralization brought about. Several regions of Europe (Spain, Austria, France) moved in that direction, as their capital cities grew out of all proportion, becoming large, unproductive centers of consumption and inhibiting the growth of their potential urban rivals. Those nations which united in their central city the dual function of national capital and maritime gateway were better able to maintain their autocatalytic dynamics. Such was the case, in the sixteenth to eighteenth centuries, of Britain and the United Provinces. Like older cores of the Network (Venice, Genoa, Antwerp) London and Amsterdam were maritime cities, and constant contact with the sea (more than any specifically English or Dutch cultural trait) inspired and sustained their elites' outward orientation. A similar effect might have been achieved in Spain and even in China:

> When in 1421 the Ming rulers of China changed their capital city, leaving Nanking, and moving to Peking...the massive world-economy of China swung round for good, turning its back on a form of economic activity based on easy access to sea-borne trade. A new landlocked metropolis was now established deep in the interior and began to draw everything towards it....

Philip II made an equally momentous decision in 1582. At the height of Spain's political domination of Europe, Philip II conquered Portugal and elected residence, with his government, in Lisbon for a period of almost three years.... Looking over the ocean this was an ideal place from where to rule the world.... So to leave Lisbon in 1582 meant leaving a position from which the empire's entire economy could be controlled, and imprisoning the might of Spain in Madrid, the landlocked heart of Castile—a fateful mistake! The Invincible Armada, after years of preparation, sailed to its disaster in 1588.[74]

Although most European and non-European elites were very aware of the importance of sea power and of the profits of long-distance trade, only constant contact with the sea seems to have convinced them to partake of the colossal benefits inherent in the energy trapped in winds and currents. The oceans and the atmosphere form a nonlinear dynamical system that contains ten times more solar energy than plants capture through photosynthesis, and only a tiny fraction of the potential energy of plant life powered most of civilization's past intensifications. The enormous reservoir of oceanic and atmospheric energy fuels a great variety of self-organized structures: tornadoes, cyclones, pressure blocks, and, more importantly for human history, wind circuits.

Some of these circuits, like the monsoon wind, which has powered all sail ships in Asian waters for centuries, gave societies a clock, a periodical rhythm. The monsoon blows westward half the year and eastward the other half, creating a "seasonal weather system that could be comprehended from land,"[75] and could thus enter as a factor in the decision-making processes of the seafaring towns in Asia. In those urban centers in contact with the monsoon, knowledge of its dynamical behavior accumulated and skills in the art of tapping its energy with sails developed. Similar knowledge and skills evolved in the ports and metropolitan centers on the Mediterranean. However, these skills were inadequate to master the circuit that would change the course of the millennium: the gigantic "double conveyor belt" formed by the trade winds and the westerlies, the wind circuit that brought Europeans to the New World and back again. Harnessing the energy of this conveyor belt, which allowed the conversion of an entire continent into a rich supply zone to fuel the growth of European cities, required special skills, and these had accumulated by the fifteenth century in European cities facing the Atlantic, particularly in Lisbon.

In the expanse of water between the Iberian peninsula and the Canary Islands, a small-scale replica of this double conveyor belt existed. The trip from Europe to the islands was straightforward, but the return was difficult since it was against the wind. The solution was to navigate away from that wind—something that sailors from Mediterranean or Indian Ocean ports would never try—and look for another one which blew in the opposite direction. This strategy of using two different circuits, one to go and one to come back, was developed by the sailors of Lisbon, and called *volta do mar*. It was later adapted by a native of Genoa in his effort to discover a western route to the Orient:

> The alternating use of the trade winds on the outward leg, then the volta (the crabwise slide off to the northwest) to the zone of the westerlies, and then to swoop home with the westerlies as the following winds ... made the gambles of Columbus, da Gama and Magellan acts of adventure not acts of probable suicide. The sailors knew they could sail out on the trades and back on the westerlies.... It is doubtful if the sailors of the age of exploration thought of the volta in any sort of formal way. It is improbable that they learned the technique as a principle; they were, after all, groping out to the sea for a favorable wind not searching for laws of nature. But prevailing patterns of thought grew up to match the patterns of prevailing winds, and Iberian sailors used the volta as a template with which to plot their courses to Asia, to the Americas and around the world.[76]

Day-to-day contact with the small-scale version of the double conveyor belt generated the skills that—in combination with the growing reservoirs of human capital in these gateway cities—allowed the mastery of the Atlantic sea routes. As this knowledge spread to other metropolises, the nations that would eventually emerge and dominate the next five hundred years would be the ones that incorporated these outward-oriented cities and used them as internal motors. Those nations whose capitals were landlocked became victims of the extreme viscosity of land transport and of the tyranny of distance and its consequent hierarchical urban patterns. The story was the exact opposite for gateway cities:

> Although the conquerors, traders, and settlers planted the flag of their sovereign, a limited number of ports actually directed the expansion. [Gateway] cities developed ties to overseas settlements and to one another that were stronger than their links with the territory at their back. As a group, they constituted the core of a powerful trading network whose outposts spanned the world and through which, via overseas gateways, were funneled the plunder and produce of vast regions.[77]

Despite the fact that the analysis of urban dynamics which I have attempted here is merely a sketch, ignoring so many other important historical factors affecting cities, it nevertheless provides certain insights into the role nonlinear science might play in the study of human history. First and foremost, nonlinear models show that without an energy flow of a certain intensity, no system, whether natural or cultural, can gain access to the self-organization resources constituted by endogenously generated stable states (attractors) and transitions between those states (bifurcations). Second, nonlinear models illustrate how the structures generated by matter-energy flows, once in place, react back on those flows either to inhibit them or further intensify them. We have seen that many different types of structures can play this catalytic role: the mineralized infrastructure of cities themselves; the organizations (centralized or decentralized) that live within the mineral walls; and various other cultural materials that move in and out of cities or accumulate in them: skills and knowledge, money and credit, informal rules and institutional norms. Furthermore, wars and antimarket rivalries between cities (and, later on, nation-states) also had catalytic effects on all these flows.[78] It was precisely these catalysts acting on each other (in autocatalytic or cross-catalytic relations), in the context of an intensified energy flow, that propelled Europe ahead of its potential rivals for world domination.

To the extent that these basic insights are correct, human culture and society (considered as dynamical systems) are no different from the self-organized processes that inhabit the atmosphere and hydrosphere (wind circuits, hurricanes), or, for that matter, no different from lavas and magmas, which as self-assembled conveyor belts drive plate tectonics and over millennia have created all the geological features that have influenced human history. From the point of view of energetic and catalytic flows, human societies are very much like lava flows; and human-made structures (mineralized cities and institutions) are very much like mountains and rocks: accumulations of materials hardened and shaped by historical processes. (There are, of course, several ways in which we are *not* like lava and magma, and these differences will be discussed in the following chapters.)

Meanwhile, this "geological" approach to human history still has some surprises in store for us as we explore the last three hundred years of the millennium. During those centuries, the population of towns which had propelled Europe into her position of worldwide supremacy witnessed dramatic changes. Just as powerful intensifications of the flows of energy had triggered the great acceleration of city building between the years 1000 and 1300, fossil fuels would make a new round of intensified energy

flow possible five centuries later and would dramatically alter the composition of this population, accelerating city births once more and giving
rise to novel forms, such as the factory town completely controlled by its
industrial hierarchies: a truly mineralized antimarket.

Sandstone and Granite

The concepts of "meshwork" and "hierarchy" have figured so prominently in our discussion up to this point that it is necessary to pause for a moment and reflect on some of the philosophical questions they raise. Specifically, I have applied these terms in such a wide variety of contexts that we may very well ask ourselves whether some (or most) of these applications have been purely metaphorical. There is, no doubt, some element of metaphor in my use of the terms, but there are, I believe, common physical processes behind the formation of mesh-works and hierarchies which make each different usage of the terms quite literal. These common processes cannot be fully captured through linguistic representations alone; we need to employ something along the lines of *engineering diagrams* to specify them.

A concrete example may help clarify this crucial point. When we say (as marxists used to say) that "class struggle is the motor of history" we are using the word "motor" in a purely metaphorical sense. However, when we say that "a hurricane is a steam motor" we are *not* simply making a linguistic analogy; rather, we are saying that hurricanes embody the same diagram used by engineers to *build* steam motors—that is, we are saying that a hurricane, like a steam engine, contains a reservoir of heat, operates via thermal differences, and circulates energy and materials through a (so-called) Carnot cycle.[79] (Of course, we may be wrong in ascribing this diagram to a hurricane, and further empirical

research may reveal that hurricanes in fact
operate in a different way, according to a different diagram.)

I wish to argue here that there are also
abstract machines (as Deleuze and Guattari
call these engineering diagrams) behind the
structure-generating processes that yield as
historical products specific meshworks and
hierarchies. Particularly instructive among
hierarchical structures are social strata
(classes, castes). The term "social stratum"
is itself clearly a metaphor, involving the
idea that, just as geological strata are layers
of rocky materials stacked on top of each
other, so classes and castes are layers—
some higher, some lower—of human materials. Is it possible to go beyond metaphor
and show that the genesis of both geological
and social strata involves the same engineering diagram? Geological strata are created
by means of (at least) two distinct operations.
When one looks closely at the layers of rock
in an exposed mountainside, one is struck by
the observation that each layer contains further layers, each composed of pebbles that
are nearly *homogeneous* with respect to size,
shape, and chemical composition. Since

pebbles do not come in standard sizes and shapes, some kind of *sorting mechanism* must be involved here, some specific device to take a multiplicity of pebbles of heterogeneous qualities and distribute them into more or less uniform layers.

Geologists have discovered one such mechanism: rivers acting as veritable *hydraulic computers* (or, at least, sorting machines). Rivers transport rocky materials from their point of origin (an eroding mountain) to the bottom of the ocean, where these materials accumulate. In the course of this process, pebbles of various size, weight, and shape react differently to the water transporting them. Some are so small they dissolve in the water; some are larger and are carried in suspension; even larger stones move by jumping back and forth from the riverbed to the streaming water, while the largest ones are moved by traction as they roll along the bottom toward their destination. The intensity of the river flow (i.e., its speed and other intensities, such as temperature or clay saturation) also determines the outcome, since a large pebble that could only be rolled by a moderate current may be transported in suspension by a powerful eddy. (Since there is feedback between pebble properties and flow properties, as well as between the river and its bed, the "sorting computer" is clearly a highly nonlinear dynamical system.)[80]

Once the raw materials have been sorted out into more or less homogeneous groupings deposited at the bottom of the sea (that is, once they have become sedimented), a second operation is necessary to transform these loose collections of pebbles into a larger-scale entity: sedimentary rock. This operation consists in *cementing* the sorted components together into a new entity with emergent properties of its own, that is, properties such as overall strength and permeability which cannot be ascribed to the sum of the individual pebbles. This second operation is carried out by certain substances dissolved in water (such as silica or hematite, in the case of sandstones) which penetrate the sediment through the pores between pebbles. As this percolating solution crystallizes, it *consolidates* the pebbles' temporary spatial relations into a more or less permanent "architectonic" structure.[81]

Thus, a double operation, a "double articulation" transforms structures on one scale into structures on another scale. In the model proposed by Deleuze and Guattari, these two operations constitute an engineering diagram and so we can expect to find isomorphic processes (that is, this same "abstract machine of stratification") not only in the world of geology but in the organic and human worlds as well.[82] For example, according to neo-Darwinians, species form through the slow accumulation of genetic materials and the adaptive anatomical and behavioral traits that those genetic materials yield when combined with nonlinear dynamical processes (such as the interaction of cells during the development of an embryo). Genes, of course, do not merely deposit at random but are sorted out by a variety of selection pressures, including climate, the action of predators and parasites, and the effects of male or female choice during mating. Thus, in a very real sense, genetic materials "sediment" just as pebbles do, even if the nonlinear dynamical system that performs the sorting operation is completely different in detail. Furthermore, these loose collections of genes can (like accumulated sand) be lost under drastically changed conditions (such as the onset of an ice age) unless they consolidate. This second operation is performed by "reproductive isolation": when a given subset of a population becomes mechanically or genetically incapable of mating with the rest. Reproductive isolation acts as a "ratchet mechanism" that conserves the accumulated adaptation and makes it impossible for a given population to "de-evolve" all the way back to unicellular organisms. Through selective accumulation and isolative consolidation, individual animals and plants come to form a larger-scale entity: a new species.[83]

We also find these two operations (and hence, this abstract diagram) in the formation of social classes. We talk of "social strata" whenever a given society presents a variety of differentiated roles to which individuals are denied equal access, and when a subset of those roles (to which a ruling elite alone has access) involves the control of key energy and material resources. While role differentiation may be a spontaneous effect of an intensification in the flow of energy through society (e.g., when a Big Man in prestate societies acts as an *intensifier* of agricultural production[84]), the sorting of those roles into ranks on a scale of prestige involves specific group dynamics. In one model, for instance, members of a group who have acquired preferential access to some roles begin to acquire the power to control further access to them, and within these dominant groups criteria for sorting the rest of society into subgroups begin to crystallize.[85]

Even though most cultures develop some rankings of this type, not in all societies do these rankings become an *autonomous dimension* of social organization. In many societies differentiation of the elites is not extensive (they do not form a center while the rest of the population forms an excluded periphery), surpluses do not accumulate (they may, for instance, be destroyed in ritual feasts), and primordial relations (of kin and local alliances) tend to prevail. Hence, for social classes or castes to become a separate entity, a second operation is necessary beyond the mere sorting of people into ranks: the informal sorting criteria need to be given a theological interpretation and a legal definition, and the elites need to become the guardians and bearers of the newly institutionalized tradition, that is, the legitimizers of change and delineators of the limits of innovation. In short, to transform a loose ranked accumulation of traditional roles (and criteria of access to those roles) into a social class, the latter needs to become consolidated via theological and legal codification.[86]

No doubt, this characterization of the process through which social strata emerge is somewhat simplified; even geological strata are more complicated than this. (For example, they grow not only through sedimentation but also through accretion and encroachment. Species and social classes may also involve these mechanisms.) But I will retain here the simplified diagram for its heuristic value: sedimentary rocks, species, and social classes (and other institutionalized hierarchies) are all historical constructions, the product of definite structure-generating processes that take as their starting point a heterogeneous collection of raw materials (pebbles, genes, roles), homogenize them through a sorting operation, and then consolidate the resulting uniform groupings into a more permanent state. The hierarchies to which I have referred throughout this chapter are a special case of a more general class of structures, stratified systems, to which not only human bureaucracies and biological species belong, but also sedimentary rocks. (And all this *without metaphor*.)

What about meshworks? Deleuze and Guattari offer a hypothetical diagram for this type of structure, too, but its elements are not as straightforward as those involved in the formation of strata. Perhaps the most-studied type of meshwork is the "autocatalytic loop," a closed chain of chemical processes, which must be distinguished from the simple self-stimulating dynamics to which I referred many times in my description of turbulent urban growth. Unlike simple autocatalysis, a closed loop displays not only self-stimulation but also self-maintenance; that is, it links a series of mutually stimulating pairs into a structure that reproduces as a whole.

The physical basis for either simple or complex self-stimulation are catalysts, that is, chemical substances capable of "recognizing" a more or less specific material and altering that material's molecular state so that it now reacts with certain substances with which it would not normally react. This act of recognition is not, of course, a cognitive act but one effected through a lock-and-key mechanism: a portion of the catalytic molecule fits or meshes with a portion of the target molecule, changing its internal structure so that it becomes more or less receptive to yet another substance. In this way, the catalyst *provokes a meeting of two substances*, facilitating (or inhibiting) their reaction and, therefore, the accumulation (or decumulation) of the products of that reaction. Under special conditions, a set of these processes may form a closed loop, where the product that accumulates due to the acceleration of one reaction serves as the catalyst for yet another reaction, which in turn generates a product that catalyzes the first one. Hence, the loop becomes self-sustaining for as long as its environment contains enough raw materials for the chemical reactions to proceed.

Humberto Maturana and Francisco Varela, pioneers in the study of autocatalytic loops, distinguish two general characteristics of these closed circuits: they are dynamical systems that endogenously generate their own *stable states* (called "attractors" or "eigenstates"), and they grow and evolve by *drift*.[87] The first characteristic may be observed in certain chemical reactions involving autocatalysis (as well as cross-catalysis) which function as veritable "chemical clocks," that is, the accumulations of materials from the reactions alternate at *perfectly regular intervals*. If we imagine each of the two substances involved as having a definite color (say, red and blue), their combination would not result in a purple liquid (as we would expect from millions of molecules combining at random) but in a rhythmic reaction with states in which mostly blue molecules accumulate followed by states in which mostly red molecules are produced. This rhythmic behavior is not imposed on the system from the outside but generated spontaneously from within (via an attractor).[88]

The second characteristic mentioned by Maturana and Varela, growth by drift, may be explained as follows: in the simplest autocatalytic loops there are only two reactions, each producing a catalyst for the other. But once this basic two-node network establishes itself, new nodes may insert themselves into the mesh as long as they do not jeopardize its internal consistency. Thus, a new chemical reaction may appear (using previously neglected raw materials or even waste products from the original loop) that catalyzes one of the original reactions and is catalyzed by the other, so that the loop now becomes a three-node network. The meshwork has now grown, but in a direction that is, for all practical purposes, "unplanned." A new node (which just happens to satisfy some internal consistency requirements) is added and the loop complexifies, yet precisely because the only constraints were internal, the complexification does not take place *in order for* the loop as a whole to meet some external demand (such as adapting to a specific situation). The surrounding environment, as source of raw materials, certainly constrains the growth of the meshwork, but more in a proscriptive way (what not to do) than in a prescriptive one (what to do).[89]

The question now is whether we can derive from empirical studies of meshwork behavior a structure-generating process that is abstract enough to operate in the worlds of geology, biology, and human society. In the model proposed by Deleuze and Guattari, there are three elements in this diagram. First, a set of heterogeneous elements is brought together via an *articulation of superpositions*, that is, an interconnection of diverse but overlapping elements. (In the case of autocatalytic loops, the nodes in the circuit are joined to each other by their *functional complementarities*.) Second, a special class of operators, or *intercalary elements*, is needed to effect these interconnections. (In our case, this is the role played by catalysts, which insert themselves between two other chemical substances to facilitate their interaction.) Finally, the interlocked heterogeneities must be capable of endogenously generating stable patterns of behavior (for example, patterns at regular temporal or spatial intervals).[90] Is it possible to find instances of these three elements in geological, biological, and social structures?

Igneous rocks (such as granite) are formed in a process radically different from sedimentation. Granite forms directly out of cooling magma, a viscous fluid composed of a diversity of molten materials. Each of these liquid components has a different threshold of crystallization; that is, each undergoes the bifurcation toward its solid state at a different critical point in temperature. As the magma cools down, its different elements separate as they crystallize in sequence, and those that solidify earlier serve as containers for those that acquire a crystal form later. The result is a complex set of heterogeneous crystals that *interlock* with one another, and this is what gives granite its superior strength.[91]

The second element in the diagram, intercalary operators, includes, in addition to catalytic substances, anything that brings about local articulations from within—"densifications, intensifications, reinforcements, injections, showerings, like so many intercalary events."[92] The reactions between liquid magma and the walls of an already crystallized component, nucleation events within the liquid which initiate the next crystallization, and even certain "defects" inside the crystals (called "dislocations") which promote growth from within, are all examples of intercalary elements. Finally, some chemical reactions within the magma may also generate endogenous stable states. When a reaction like the one involved in chemical clocks is not stirred, the temporal intervals generated become spatial intervals, forming beautiful spiral and concentric-circle patterns that can be observed in frozen form in some igneous rocks.[93]

Thus, granite (as much as a fully formed autocatalytic loop) is an instance of a meshwork, or, in the terms used by Deleuze and Guattari, a *self-consistent aggregate.* Unlike Maturana and Varela, who hold that the quality of self-consistency exists only in the biological and linguistic worlds,

Deleuze and Guattari argue that "consistency, far from being restricted to complex life forms, fully pertains even to the most elementary atoms and particles."[94] Therefore we may say that much as hierarchies (organic or social) are special cases of a more abstract class, strata, so autocatalytic loops are special cases of self-consistent aggregates. And much as strata are defined as an articulation of homogeneous elements, which neither excludes nor requires the specific features of hierarchies (such as having a chain of command), so self-consistent aggregates are defined by their articulation of heterogeneous elements, which neither excludes nor requires the specific features of autocatalytic loops (such as growth by drift or internal autonomy). Let's now give some biological and cultural examples of the way in which the diverse may be articulated as such via self-consistency.

A species (or more precisely, the gene pool of a species) is a prime example of an organic stratified structure. Similarly, an ecosystem represents the biological realization of a self-consistent aggregate. While a species may be a very homogeneous structure (especially if selection pressures have driven many genes to fixation), an ecosystem links together a wide variety of heterogeneous elements (animals and plants of different species), which are articulated through interlock, that is, by their functional complementarities. Given that the main feature of an ecosystem is the circulation of energy and matter in the form of food, the complementarities in question are alimentary: prey-predator or parasite-host are two of the most common functional couplings in food webs. *Symbiotic relations* can act as intercalary elements, aiding the process of building food webs (an obvious example: the bacteria that live in the guts of many animals, which allows those animals to digest their food).[95] Since food webs also produce endogenously generated stable states, all three components of the abstract diagram would seem to be realized in this example.[96]

We have already observed several examples of cultural meshworks which also fit our description of self-consistent aggregates. The simplest case is that of small-town markets. In many cultures, weekly markets have been the traditional meeting place for people with heterogeneous needs. Matching, or interlocking, people with complementary needs and demands is an operation that is performed automatically by the price mechanism. (Prices transmit information about the relative monetary value of different products and create incentives to buy and sell.) As Herbert Simon observes, this interlocking of producers and consumers could in principle be performed by a hierarchy, but markets "avoid placing on a central planning mechanism a burden of calculation that such a mechanism, however well buttressed by the largest computers, could not sustain. [Markets] conserve information and calculation by making it possible to assign decisions to the actors who are most likely to possess the information (most of it local in origin) that is relevant to those decisions."[97]

Of course, for this mechanism to work prices must *set themselves*, and therefore we must imagine that there is *not* a wholesaler in town who can manipulate prices by dumping large amounts of a given product into the market (or by hoarding). In the absence of price manipulation, money (even primitive forms of money, such as salt, shells, or cigarettes) functions as an intercalary element: with pure barter, the possibility of two exactly matching demands meeting by chance is very low; with money, those chance encounters become unnecessary and complementary demands may find each other at a distance, so to speak. Other intercalary elements are also needed to make markets work. As we have repeatedly noted, not just material and energetic resources change hands in a market, property rights (the legal rights to use those resources) do too. Hence we typically do not have to model simple exchanges but more complex transactions that involve a host of other costs, such as those involved in enforcing agreements. If these transaction costs are too high, the gains from trade may evaporate. In small-town markets, informal constraints (such as codes of behavior enforced through peer pressure in dense social networks) are also needed to reduce transaction costs and allow the interlocking of complementary demands to take place.[98] Finally, markets also seem to generate endogenous stable states, particularly when commercial towns form trading circuits, as can be seen in the cyclical behavior of their prices, and this provides us with the third element of the diagram.

Thus, much as sedimentary rocks, biological species, and social hierarchies are all stratified systems (that is, they are each the historical product of a process of double articulation), so igneous rocks, ecosystems, and markets are self-consistent aggregates, the result of the coming together and interlocking of heterogeneous elements. And just as the diagram defining the "stratifying abstract machine" may turn out to require more complexity than our basic diagram of a double articulation, so we may one day discover (empirically or through theorizing and computer simulations) that the diagram for the meshwork-producing process involves more than the three elements outlined above. Moreover, in reality we will always find mixtures of markets and hierarchies, of strata and self-consistent aggregates. As Simon says, it may seem prima facie correct to say that whereas markets figure most prominently in coordinating economic activities in capitalist countries, hierarchic organizations play the largest role in socialist countries. But that is too simple a formula to describe the realities which always exhibit a blend of all the mechanisms of coordination. The economic units in capitalist societies are mostly business firms, which are themselves hierarchic organizations, some of enormous size, that make only a modest use of markets in their internal functioning. Conversely socialist states use market prices to a growing extent to supplement hierarchic control in achieving inter-industry coordination.[99]

There is one final aspect of meshwork dynamics I must examine before returning to our exploration of the "geological" history of human societies. We may wonder why, given the ubiquity of self-consistent aggregates, it seems so hard to think about the structures that populate the world in any but hierarchical terms. One possible answer is that stratified structures involve the simplest form of *causal relations*, simple arrows going from cause to effect.[100] According to Magoroh Maruyana, a pioneer in the study of feedback, Western thought has been dominated by notions of linear (nonreciprocal) causality for twenty-five hundred years. It was not until World War II that the work of Norman Wiener (and engineers involved in developing radar systems) gave rise to the study of negative feedback and with it the beginning of nonlinear thinking.

The classic example of negative feedback is the thermostat. A thermostat consists of at least two elements: a sensor, which detects changes in ambient temperature, and, an effector, a device capable of changing the ambient temperature. The two elements are coupled in such a way that whenever the sensor detects a change beyond a certain threshold it causes the effector to modify the surrounding temperature in the opposite direction. The cause-and-effect relation, however, is not linear (from sensor to effector) since the moment the effector causes a change in the surrounding temperature it thereby affects the subsequent behavior of the sensor. In short, the causal relation does not form a straight arrow but folds back on itself, forming a closed loop. The overall result of this circular causality is that ambient temperature is maintained at a given level.

Maruyana opposes negative feedback with "positive feedback" (a form of nonlinear causality that we have already encountered in the form of autocatalysis). While the first type of reciprocal causality was incorporated into Western thought in the 1950s, the second type had to wait another decade for researchers like Stanislav Ulam, Heinz Von Foerster, and Maruyana himself to formalize and develop the concept.[101] The turbulent dynamics behind an explosion are the clearest example of a sys-

tem governed by positive feedback. In this case the causal loop is established between the explosive substance and its temperature. The velocity
of an explosion is often determined by the intensity of its temperature
(the hotter the faster), but because the explosion itself generates heat,
the process is self-accelerating. Unlike the thermostat, where the arrangement helps to keep temperature under control, here positive feedback
forces temperature to go out of control. Perhaps because positive feedback is seen as a destabilizing force many observers have tended to
undervalue it relative to negative feedback. (In the so-called Gaia hypothesis, for instance, where stabilizing negative feedback is postulated to
exist between living creatures and their environment, positive feedback is
sometimes referred to pejoratively as "anti-Gaian.")[102]

Maruyama sees the question in different terms. For him the principal
characteristic of negative feedback is its homogenizing effect: any deviation from the temperature threshold at which the thermostat is set is
eliminated by the loop. Negative feedback is "deviation-counteracting."
Positive feedback, on the other hand, tends to increase heterogeneity by
being "deviation-amplifying": two explosions set off under slightly different conditions will arrive at very different end states, as the small original
differences are amplified by the loop into large discrepancies.[103] We have
already observed the many roles that positive feedback has played in the
turbulent history of Western towns. However, it is important to distinguish
between simple autocatalytic dynamics and complex autocatalytic loops,
which involve not only self-stimulation but self-maintenance (that is, positive feedback and closure).

Another way of stating this distinction is to say that the increase in
diversity that mutually stimulating loops bring about will be short-lived unless the heterogeneous elements are interwoven together, that is, unless
they come to form a meshwork. As Maruyama writes, "There are two ways
that heterogeneity may proceed: through *localization* and through *interweaving*. In localization the heterogeneity between localities increases, while
each locality may remain or become homogenous. In interweaving, heterogeneity in each locality increases, while the difference between localities
decreases."[104] In other words, the danger with positive feedback is that
the mere production of heterogeneity may result in isolationism (a high
diversity of small cliques, each internally homogeneous). Hence the need
for intercalary elements to aid in articulating this diversity without homogenization (what Maruyama calls "symbiotization of cultural heterogeneity").

Negative feedback, as a system of control and reduction of deviation,
may be applied to human hierarchies. Decision making in stratified social
structures does not always proceed via goal-directed analytic planning but

often incorporates automatic mechanisms of control similar to a thermo-stat (or any other device capable of generating homeostasis).[105] On the other hand, social meshworks (such as the symbiotic nets of producers whom Jacobs describes as engaged in volatile trade) may be modeled on positive-feedback loops as long as our model also incorporates a means for the resulting heterogeneity to be interwoven. Moreover, specific insti-tutions will likely be mixtures of both types of reciprocal causality, and the mixtures will change over time, allowing negative or positive feedback to dominate at a given moment.[106] The question of mixtures should be also kept in mind when we judge the relative *ethical value* of these two types of structure. If this book displays a clear bias against large, centralized hier-archies, it is only because the last three hundred years have witnessed an excessive accumulation of stratified systems at the expense of mesh-works. The degree of homogeneity in the world has greatly increased, while heterogeneity has come to be seen as almost pathological, or at least as a problem that must be eliminated. Under the circumstances, a call for a more decentralized way of organizing human societies seems to recom-mend itself.

However, it is crucial to avoid the facile conclusion that meshworks are *intrinsically better* than hierarchies (in some transcendental sense). It is true that some of the characteristics of meshworks (particularly their resilience and adaptability) make them desirable, but that is equally true of certain characteristics of hierarchies (for example, their goal-directed-ness). Therefore, it is crucial to avoid the temptation of cooking up a narrative of human history in which meshworks appear as heroes and hierarchies as villains. Not only do meshworks have dynamical properties that do not necessarily benefit humanity (for example, they grow and develop by drift, and that drift need not follow a direction consistent with a society's values), but they may contain heterogeneous components that are themselves inconsistent with a society's values (for example, cer-tain meshworks of hierarchies). Assuming that humanity could one day agree on a set of values (or rather on a way of meshing a heterogeneous collection of partially divergent values), further ethical judgments could be made about specific mixtures of centralized and decentralized compo-nents in specific contexts, but never about the two pure cases in isolation.

The combinatorial possibilities—the number of possible hybrids of meshworks and hierarchies—are immense (in a precise technical sense),[107] and so an *experimental and empirical* attitude toward the problem would seem to be called for. It is surely impossible to determine *purely theoreti-cally* the relative merits of these diverse combinations. Rather, in our search for viable hybrids we must look for inspiration in as many domains as possible. Here, we have looked to a realm that would normally seem out of bounds: the mineral world. But in a nonlinear world in which the same basic processes of self-organization take place in the mineral, organic, and cultural spheres, perhaps rocks hold some of the keys to understanding sedimentary humanity, igneous humanity, and all their mixtures.

Geological History: 1700–2000 A.D.

Prior to the eighteenth century all the energetic intensifications that humanity had undertaken were relatively short-lived. The intensified exploitations of agricultural resources which had sustained wave after wave of ancient urbanization were typically followed by soil depletion or erosion, bringing human expansion to a halt. Even the more recent acceleration of city building in Europe at the turn of the millennium, which added commercial and proto-industrial positive feedback to the process, was followed by a long depression. The first intensification to escape this cyclical destiny, beginning roughly in the year 1700, was based on the burning of energy-rich ore. Coal is the product of one of several types of mineralization that organic matter can undergo. When the corpses of plants and animals accumulate under water in the absence of oxygen, the microorganisms that would normally remineralize them and recycle them in the ecosystem cannot operate; hence these deposits do not rot. Instead, they are compressed, carbon-enriched, and eventually petrified. Although several ancient societies had made use of these rocks, England was the first civilization to submit coal deposits to intense exploitation, creating the principal flow of nonhuman energy with which to fuel its industrial revolution.

This new intensification had dramatic consequences for the population of towns and cities of Europe, as well as for the institutions that inhabited them. Here we will examine several of these consequences, taking advantage of the novel insights on the origins and dynamics of the Industrial Revolution proffered by historians and theorists who have applied to their subject concepts borrowed from nonlinear science. In particular, here the rise of the "industrial age" will not be viewed as the result of human society having reached a new "stage of development" (a new mode of production) or of its having climbed further up the ladder of progress, but, rather, as the crossing of a bifurcation where previous autocatalytic dynamics (subject to negative feedback) came to form a self-sustaining autocatalytic loop.

Moreover, technology won't be viewed as evolving in a straight line, as if the advent of steam power and factory production were the inevitable outcome of the evolution of machines. On the contrary, mass production techniques in all their forms were only one alternative among several, and the fact that they came to dominate the development of new machinery is itself in need of explanation. Our investigation of the intensifications that fossil fuels made possible begins with steam power and moves on to electricity, which formed the basis for a second industrial revolution in our own century. Both coal and steam, and later oil and electricity, greatly affected the further development of Western towns, and, as usual, once the mineralized infrastructure of those towns, and the institutions within them, had registered the effects of these intensifications, they reacted back on the energy flows to constrain them, either inhibiting them or further intensifying them.

Although Europe underwent a long period of relatively slow economic growth after 1300, the population of European towns nonetheless underwent significant change. The long depression had acted like a "sorting device," eliminating many towns on the lower ranks of Central Place hierarchies and concentrating growth at the top. Consequently, the command element in the mix had increased (as had its degree of homogenization, due to the absorption of cities and their regions into nation-states). The relatively few new European cities that were born between 1300 and 1800 were planned cities (usually port cities created by central governments in order to enter the great maritime races). For example, between 1660 and 1715, the French hierarchies under Louis XIV created a strategic network of commercial and military port cities—Brest, Lorient, Rochefort, and Sète—each one planned "to play a specific role in the government's politico-military strategy for sea-power."[108]

By contrast, in the 1800s the intense circulation of coal energy gave rise to a far greater number of new (mining and factory) towns, most of which grew spontaneously, not to say chaotically. This was the case, for example, in the Ruhr region, which would later become the center of German heavy industry, and in Lancashire, the heart of industrial Britain. In these two regions, and others, mills, mining centers, and metallurgical complexes mushroomed everywhere, unregulated and having no systematic relations with one another. Some older cities, such as Liverpool and Manchester, grew enormously (one becoming the gateway, the other the capital of the region), while a multitude of new towns sprang up around them: Bolton, Bury, Stockport, Preston, Blackburn, Burnley. As these coal-fueled towns devoured the countryside and grew into each other, they formed huge *conurbations*: extremely dense but weakly centralized urban regions produced by accelerated industrialization. In the words of Hohenberg and Lees:

> The best examples of the transforming power of rapid industrial growth are to be found in the coal-mining regions. There the explosive concentrated effects of . . . modern economic change can be seen in pure form. Since coal was needed to run the engines and smelt the ores, factories and fur-

naces tended to locate very near coal supplies or in places where they had good access to transportation. As demand skyrocketed, mining areas with their expanding number of pits, workshops and new firms attracted new workers.... Both high fertility and migration bred an extreme density of settlement, which soon surpassed anything that the proto-industrial era had known. These coal basins grew by a kind of regional implosion, whereby a rural milieu crystallized into a densely urban one.[109]

These new towns would soon be inhabited by an industry that was un-doubtedly more complex than anything humanity had seen before. And yet, as Hohenberg and Lees remind us, it was not as if society as a whole had reached a new stage and every region now moved in lockstep toward this type of industrialization. Not only were there regions that industrialized in a different way, but still others underwent radical deindustrialization. Industrial development is like biological evolution, which not only lacks any progressive direction, it does not even have a consistent drive toward complexification: while some species complexify, others simplify.[110]

In both cases, a variety of processes result in accumulations of com-plexity in some areas, deaccumulations in others, and the coexistence of different types of accumulated complexity. The large-scale, concentrated industry of coal-fueled towns represented only one possible direction for the complexification of technology. Areas that industrialized more slowly and maintained their ties to traditional craft skills developed methods of production that were scattered and small in scale but highly sophisti-cated, with a complex division of labor and a high degree of market in-volvement. "Whether one looks at Swiss cottons and watches, at textiles in Piedmont and the Vosges, or at metalwares in central Germany the pic-ture is the same: upland valleys fashioning an enduring industrial posi-tion without ever turning their backs on the proto-industrial heritage."[111]

Thus, there were at least two stable trajectories for the evolution of industry, proceeding at different speeds and intensities: large-scale, energy-intensive industry and small-scale, skill-intensive industry. While the former gave rise to functionally homogeneous towns, in many cases controlled by their industrial hierarchies (the factory town), the latter was housed in small settlements, with a more heterogeneous set of eco-nomic functions and less concentrated control. Antimarket institutions took over only one type of industry, that which, like themselves, was based on *economies of scale*.

Besides differing in the proportion of meshwork and hierarchy in their mixes, these towns also varied in terms of the form of their expansion. The rapid, violent growth of coal-fueled cities, which expanded into the countryside with total disregard for previous land-use patterns, contrasts with the way in which the small towns that housed decentralized industries meshed with their rural surroundings.[112] Although all towns tend to dominate their countrysides, industrial towns intensified this exploitation. As the biogeographer Ian G. Simmons has noted, urban economies based on coal had a host of hidden costs—from the vast amounts of diverted water they used to the depressions, cracks, and sinkholes that continued to form long after mining had stopped—and the surrounding rural areas bore the brunt of those ecological costs.[113]

Simmons views cities as veritable transformers of matter and energy: to sustain the expansion of their exoskeleton, they extract from their surroundings sand, gravel, stone, and brick, as well as the fuel needed to convert these into buildings. He notes that, like any system capable of self-organization, cities are open (or dissipative) systems, with matter-energy flowing in and out continuously. And this is all the more true for nineteenth-century industrial towns. Besides the raw materials needed to maintain their mineralization, these towns needed to input flows of iron ores, limestone, water, human labor, and coal, as well as to output other flows (solid waste, sewage, manufactured goods). Rural areas absorbed some of the noxious outputs, while the inputs began to come from farther and farther away, particularly as groups of coal towns coalesced into conurbations. These links to faraway supply regions, plus the lack of systematic relations between services and size of settlements, placed these towns within the Network system rather than within the Central Place hierarchies.[114]

What made these urban centers special, however, was not so much the matter-energy flows that traversed them, but the way in which those flows became *amplified*. Hence, argues Simmons, while coal used for iron smelting was exploited with increasing intensity since 1709, it was not until the nineteenth century, when the steam engine had matured, that industrial takeoff occurred: "A small amount of coal invested in such an engine was the catalyst for the production of energy and materials on an ever larger scale."[115] In all dissipative systems, energy must be put in before any surpluses can be taken out. Even though an industrial town had to invest more energy than previous urban centers, it extracted greater surpluses per unit of energy. Basically, it used certain flows of energy to amplify other flows.

Furthermore, these positive-feedback links between flows began to form closed circuits: antimarket money flowed into mining regions and intensified coal extraction and iron production, which triggered a flow of mechanical energy (steam), which in turn triggered a flow of cotton textiles, which created the flow of profits that financed further experimentation with coal, iron, and steam technology. These loops of triggers and flows were behind the explosive urban growth in England between 1750 and 1850. As Richard Newbold Adams puts it, "Great Britain in this era was a great expanding dissipative structure, consuming increasing amounts of energy."[116] And precisely these autocatalytic loops were what kept this self-organized structure going:

> A trigger of one energy form sets off a flow in another which, in turn, triggers a release of a flow in the first; the insertion of more parties creates a chain of trigger-flow interactions that may go in series, in parallel or both.... The trigger-flow interactions specifically create an interdependent reproduction among the participating dissipative structures. It interlocks a series of separately reproductive systems into a single, interactive reproductive system.[117]

These meshworks of mutually supporting innovations (coal-iron-steam-cotton) are well known to historians of technology.[118] They existed long before the nineteenth century (e.g., the interlocking web formed by the horseshoe, the horse harness, and triennial rotation which was behind the agricultural intensification at the turn of the millennium), and they occurred afterward, as in the meshwork of oil, electricity, steel, and synthetic materials that contributed to the second industrial revolution. Nonetheless, as important as they were, autocatalytic loops *of technologies* were not complex enough to create a self-sustained industrial take-off. Before the 1800s, as we noted, these intensifications often led to depletions of resources and diminishing returns. Negative feedback eventually checked the turbulent growth generated by positive feedback.

Braudel uses two examples of early encounters between antimarkets and industrial technology to make this point. In some Italian cities (e.g., Milan) and some German cities (e.g., Lübeck and Cologne), explosive growth occurred as early as the fifteenth century. The German mining industry in the 1470s "stimulated a whole series of innovations ... as well as the creation of machinery, on a gigantic scale for the time, to pump out water from the mines and to bring up the ore."[119] Milan, on the other hand, witnessed an extraordinary increase in textile manufacturing, with sophisticated "hydraulic machines ... to throw, spin and mill silk, with several mechanical processes and rows of spindles all turned by a single water-wheel."[120] Although simple mutually stimulating links had developed in these cities, between mining and large-scale credit, or between textile profits and commercialized agriculture, both intensifications came to a halt in a few decades.

England herself attempted an early takeoff between 1560 and 1640, at a time when, comparatively speaking, she was a rather backward industrial nation. To catch up, the British waged a campaign of industrial espionage in Italy and imported German, Dutch, and Italian craftsmen, to effect a transfer of know-how and manufacturing techniques to their island.[121] Once a skill reservoir had been formed at home, British anti-markets gave industry a much increased scale and levels of capital investment reached new peaks of intensity. Still, self-sustained growth did not occur. One possible explanation is that autocatalytic loops need to achieve a threshold of complexity before they acquire the resilience and versatility needed to overcome diminishing returns. Hence, what made nineteenth-century England a special place was the formation of a more complex, self-maintaining circuit of triggers and flows which included a number of other catalytic elements in addition to technology and big business: a national market, a stable bank and credit system, extensive long-distance trading networks, a growing agricultural sector to feed the expanding population, and, of course, the population itself, which provided raw labor and skills.

The new intensification in agriculture, which was based on simple positive feedback (between cattle raising and the crops their manure helped fertilize) but which increased in scale due to antimarket investment, played several roles in the industrial takeoff. On the one hand, it served for a long time as the principal consumer of metal tools and hence catalyzed, and was catalyzed by, the iron industry. On the other hand, the new agricultural system (which is examined in more detail in the next chapter) favored different types of soils than those used by the previous agricultural regime, and so created a large pool of unemployed farm workers, who would provide the muscular energy for the new factories.[122] Hence, agricultural regions received inputs (iron) from, and provided inputs (labor, food) to, the factory towns, and in this sense agriculture was an important node in the autocatalytic loop. The flow of labor that this node supplied, however, was to be used mostly as raw muscular energy. Skilled labor was also needed, and reservoirs of this had begun forming in the early 1700s. Indeed, the first steam engine, a water pump in a coal mine in operation by 1712, had been the product of such skilled know-how. Although its inventor, Thomas Newcomen, may have been familiar with the basic principles of steam and the vacuum, as embodied in contemporary scientific apparatuses, he put together the first engine using mostly informal knowledge.[123] Much the same can be said for the other innovations of the eighteenth century:

In erecting a machine . . . not only visual [e.g., engineering diagrams] but tactile and muscular knowledge are incorporated into the machine by the mechanics and others who use tools and skills and judgment to give life to the visions of the engineers. Those workers—machinists, millwrights, carpenters, welders, tinsmiths, electricians, riggers, and all the rest—supply all made things with a crucial component that the engineer can never fully specify. Their work involves the laying on of knowing hands. . . . The historical significance of workers' knowledge had hardly been noticed until the British economic historian John R. Harris connected it to the technological lead that Great Britain held over the Continent during the Industrial Revolution. In the seventeenth century, Britain had converted to coal as an industrial fuel [and this involved many changes.] . . . The list of changes of techniques and apparatus is very long, but these changes are unappreciated because many (probably most) of them were made by [senior skilled] workers . . . rather than by owners or the supervisors of the works. By 1710 . . . workers' growing knowledge of the techniques of coal fuel technology had already given Britain a commanding industrial lead over France and other Continental countries.[124]

These reservoirs of skilled labor were important inputs to the factory towns and hence key nodes in the loop. Skills and know-how provided what one might call "catalytic information," that is, information capable of bringing together and amplifying flows of energy and materials. This is a good argument against labor theories of value, for which a machine is nothing but the congealed muscular energy that went into its production. Strictly speaking, this would mean there is no difference between a machine that works and one that does not (or a disassembled one). As the above quote makes clear, not only is a diagram necessary (brought into the process by an engineer) but also the skilled manual knowledge needed to implement the abstract diagram. In short, the energetic inputs to large-scale production processes required complementary inputs of catalytic information in order for the Industrial Revolution to become a self-sustaining process.

Of course, in addition to these reservoirs of factory inputs, the loop required nodes capable of absorbing the industrial output. In other words, the huge outputs of factory towns, their continuous flows of manufactured products, needed domestic and foreign markets of a *sufficient scale* to absorb them. These markets were not the product of industrial towns but of the cities that nation-states had absorbed as political capitals and gateways to the now globalized networks of exchange. Unlike local and regional markets, national markets were not the product of a process of

self-organization but of deliberate planning by a country's elites, a conscious policy known as *mercantilism*.[125] It involved not only the removal of
internal tolls and tariffs, but the construction of a communications network (roads, canals, mails) to allow commands (and traded goods) from
the capital to reach the whole country. In addition to a nationwide market,
an intensification of foreign trade and the proliferation of links between
gateway cities all over the globe were also necessary ingredients.

London, part political capital and part maritime metropolis, was instrumental in the creation of the British national and foreign markets. London also played a key role in the formation of a stable credit system, with
the creation in 1694 of the first central bank, the Bank of England, which
allowed tapping (via credit) the vast monetary reserves of Amsterdam.
As Braudel remarks, even though France had at the time a greater reservoir of natural resources than England, her credit (and taxation) system
was never as good: "artificial wealth" proved more powerful than natural
wealth.[126] Hence, the first autocatalytic loop to achieve self-sustaining
growth involved more than industrial elites. Financial and commercial
antimarkets were also key ingredients, as was the nation-state. And while
each separate elite did exercise centralized control over a given process
(the logistics of factory towns, the creation of the national market), the
revolution as a whole was the result of a true meshwork of hierarchical
structures, growing, like many meshworks, by drift:

> Can we really be satisfied with this image of a smoothly coordinated and
> evenly developing combination of sectors, capable between them of
> providing all the interconnected elements of the industrial revolution and
> meeting demands from other sectors? It conveys the misleading vision of
> the industrial revolution as a consciously pursued objective, as if Britain's
> society and economy had conspired to make possible the new Machine
> Age.... But this was certainly not how the English revolution developed.
> It was not moving towards any goal, rather it encountered one, as it was
> propelled along by that multitude of different currents which not only
> carried forward the industrial revolution but also spilled over into areas
> far beyond it.[127]

Thus, at least from the perspective where social dynamics are the
same as geological dynamics (that is, from the perspective of energy and
catalysis), the process of industrial takeoff may be viewed as a bifurcation,
from a state in which self-stimulating dynamics were not complex enough
to overcome diminishing returns, to a state in which the series of nodes
forming the circuit became a self-sustaining entity. The addition of new

nodes to the meshwork as it complexified did not occur according to a plan but simply following internal constraints; that is, each new node had to "mesh well" with the existing ones (i.e., catalyze and be catalyzed by existing nodes). As the "geological history" of the nineteenth century continued to unfold, the technologies that grew around the inanimate power of steam (as well as radically new ones) simply inserted themselves as further nodes in the growing autocatalytic loop. The railroad and the telegraph, for example, meshed well not only with one another (amplifying each other's strengths and compensating for certain weaknesses), they meshed well in the larger context of the circuit.

The new self-sustained intensification was made possible by elements of both the Central Place and the Network systems. Administrative centers and gateway ports joined factory towns to form the great circuit of triggers and flows. The Industrial Revolution, in turn, affected in several ways the future growth of cities. One of the revolution's intensified flows, the flow of cast iron, triggered the beginning of the metallization of the urban exoskeleton as the industrial regions of England began to use iron frames to build fireproof textile mills: first, a six-story cotton mill with iron columns was erected in Derby in 1792; then, in 1796, a cotton mill with iron beams and columns was built in Shrewsbury; by 1830, the internal iron frame was common in industrial and public buildings in England and France.[128] Next, the web of interlocking innovations that characterized this period generated a second wave of interacting technologies (the railroad and the telegraph), which had profound effects on the European urban system as a whole, changing the relative importance of the capital and the metropolis. Up to this point, land transport could not compete with the swift and flexible communications afforded by the sea. While terrestrial distances served to separate landlocked urban settlements, the open sea served to connect gateway cities. But the advent of steam-powered transportation removed these constraints, giving territorial capitals many of the advantages previously enjoyed by maritime cities.[129]

The coal regions of England were the birthplace of the first railroad system, adopting the "Rocket" locomotive invented by George Stephenson in 1829. This allowed the Liverpool and Manchester Railway to open for business in 1830.[130] Other railways began operating on the Continent a few years later, in France and Austria, but they remained experimental for at least ten years. Yet British leadership in steam-driven transport was soon surpassed by the United States, which a few decades earlier had been an English supply region. These former colonies had taken off economically in the second half of the eighteenth century, by means of the same small-scale autocatalytic process that had allowed Europe many centuries earlier to emerge from the shadow of Islam: volatile trade among backward cities engaged in import substitution.

According to Jane Jacobs, the first two American cities to begin this process were Boston and Philadelphia, one a British resource depot for timber and fish, the other supplying England with grain. While New York remained a captive market, Boston and Philadelphia were copying European products and replacing them with local ones, which they traded among themselves. While the innovations that came out of this process were small and unglamorous, and hence cannot be compared with the ones that emerged from the Industrial Revolution, what mattered was the reservoir of interlocking skills and procedures generated by import-substitution dynamics.[131] After the War of Independence, New York joined Boston and Philadelphia in developing a greater variety of manufactures, while San Francisco would, after the gold rush, become a gateway to the emerging global Network system.

The mechanics and engineers of these American cities created the technology that would by 1850 allow the U.S. railroads to surpass the British railway system in terms of mileage of wrought-iron rails. If bridges and factories in America were still being built out of timber, the transportation system of the new nation-state was undergoing an even more intense metallization than England's. More importantly, the technology developed in England (locomotives and railway construction techniques) was largely unsuitable for the long distances and difficult terrain of the United States, and so it could not simply be imported but had to develop locally in novel ways.[132] Hence the importance of the meshworks of small firms that had developed along the American eastern seaboard, whence the local engineering and entrepreneurial talent needed to develop the new machines was recruited.

There is another side to the success of American railroads (and to the future evolution of industrialization) which involved not meshworks but command hierarchies. While the technological elements of the system had been developed by civilian engineers from New York and Philadelphia, *military engineers* were instrumental in developing the bureaucratic management methods that came to characterize American railroads. In the words of the historian Charles F. O'Connell:

> As the railroads evolved and expanded, they began to exhibit structural and procedural characteristics that bore a remarkable resemblance to those of the Army. Both organizations erected complicated management hierarchies to coordinate and control a variety of functionally diverse, geographically

separated corporate activities. Both created specialized staff bureaus to provide a range of technical and logistical support services. Both divided corporate authority and responsibility between line and staff agencies and officers and then adopted elaborate written regulations that codified the relationship between them. Both established formal guidelines to govern routine activities and instituted standardized reporting and accounting procedures and forms to provide corporate headquarters with detailed financial and operational information which flowed along carefully defined lines of communication. As the railroads assumed these characteristics, they became America's first "big business."[133]

O'Connell points out that specific individuals from the U.S. Army Corps of Engineers played key roles in the building of a number of American railroads and in so doing faced managerial problems of a scale and complexity unknown to the local business community. They made strict accountability and bureaucratic hierarchy pivotal elements of a management style that would eventually filter through to other railway lines (and other industries). Although this generally unacknowledged military element of antimarket institutions is brought into high relief by the American experience in railroad management, it did not originate there. Indeed, the relationship between military and antimarket institutions is a very old one. By the sixteenth century, Venice had developed standardized procedures as part of the operation of its arsenal, the largest industrial complex in Europe at the time.[134] The armed sail ships built at the arsenal since 1328 were used by Venetian antimarkets not only to conduct their lucrative long-distance trade with the Levant, but also to maintain by military force their monopoly on that trade. In the eighteenth and nineteenth centuries, arsenals would again play a leading role in the standardization and routinization of the production process, influencing the future development of industrial antimarkets. In particular, military discipline was transferred to factories, the workers slowly de-skilled, and their activities rationalized.

Harry Braverman, a labor historian, acknowledges the role of bureaucratic and military hierarchies in the origins of the rationalization of labor: "France had a long tradition of attempting the scientific study of work, starting with Louis XIV's minister Colbert; including military engineers like Vauban and Belidor and especially Coulomb, whose physiological studies of exertion in labor are famous."[135] Indeed, the basic routines that would later evolve into mass production techniques were born in French military arsenals in the eighteenth century. These routines were later transferred to American arsenals, where they became institutionalized over the course of the nineteenth century, eventually developing into the "American system of manufacturing."

The American system was originally devised to create weapons with perfectly interchangeable parts. When artisans manufactured the different parts of a weapon by hand, the resulting heterogeneity made it impossible to supply fronts with spare parts. The new system first created a model of a particular weapon, and then the model served as a *standard* to be exactly replicated. But enforcing this standard, to ensure the homogeneity of the products, required a transfer—from the military to the factory—of the disciplinary and surveillance methods that had been used to maintain order in barracks and camps for over two centuries. In short, the American system transformed manufacturing from an open process based on flexible skills into a closed process based on fixed routines (enforceable through discipline and constant inspection):

> When labor was mechanized and divided in nineteenth-century arms factories, individual work assignments became more simplified while the overall production process became more complex. Coordinating and controlling the flow of work from one manufacturing stage to another therefore became vital and, in the eyes of factory masters, demanded closely regulated on-the-job behavior. Under these conditions the engineering of people assumed an importance equal to the engineering of materials. As conformity supplanted individuality in the workplace, craft skills became a detriment to production.[136]

Obviously, not all aspects of the rationalization of labor had a military origin. Military institutions played a key role, but industrial discipline had already developed (more or less independently) in certain anti-market enterprises, such as mines.[137] All that can be claimed is that the process of routinization of production in arsenals, mines, and civilian factories underwent a great intensification on both sides of the Atlantic, and this implied a large increase in the command element in the economic mix. But once again, despite the important consequences that the advent of rationalization had on the future of the economy, it is important to keep in mind all the coexisting processes taking place at this time so as not to reduce their heterogeneity to a single factor. In particular, routinization needs to be contrasted with the completely different process of innovation.[138] Routinization in its intensified (and consciously planned) form occurred in a fairly defined area of the European (and American) exoskeleton, away from the national and regional capitals which became centers of innovation. While the latter kept growing in diversity and economic heterogeneity during the nineteenth century, the towns, which underwent the intensified routinization of production, became ever more homogeneous:

> At the high end of the spectrum [of occupational homogeneity], we find the single-industry and "company" towns. Often associated with secret military technology in our time, the latter go back at least to the naval ports, such as Brest and Toulon, founded by Louis XIV. In the nineteenth century, single enterprises developed sizable towns or came to dominate an urban area. Port Sunlight (Lever) in England, Leverkusen (Bayer) in Germany, and Sochaux (Peugeot) in France, are examples. Entrepreneurs were motivated by the determination to exercise total control over the human as well as the technical environment. Non-basic employment was kept to a minimum because the paternalistic employer discouraged competition and "frivolity" in the provision of services.[139]

This homogenization of economic functions, which retained basic services and excluded competing industries, meant that the only positive feedback operating in these urban centers was the enormous economies of scale to which their antimarket institutions had access. By standardizing production, costs could be spread across a large number of identical products, and in this way the law of diminishing returns could be overcome. Yet, there are other possible types of positive feedback for cities and towns, other connections between efficiency and size—not the size of a homogenized enterprise and its homogeneous mass-produced products, but the size of a highly heterogeneous urban center which provides small firms with a variety of mutually stimulating links. These are not economies of scale, but *economies of agglomeration*:

> [These economies] come from the fact that the firm can find in the large city all manner of clients, services, suppliers, and employees no matter how specialized its product; this, in turn, promotes increased specialization. Surprisingly, however, economies of agglomeration encourage firms of the same line to locate close to one another, which is why names such as Harley, Fleet, and Lombard streets and Saville Row—to stick to London— call to mind professions rather than place. Besides the non-negligible profit and pleasure of shop-talk, all can share access to services that none could support alone.... A key point about economies of agglomeration is that small businesses depend on them more than do large ones. The latter can internalize these "external economies" by providing their own services and gain locational freedom as a result.... The relationship between large

cities and small business is a symbiotic one beneficial to both. The reason is that small firms are the major carriers of innovation, including creative adaptation to change. This was even more true in the days before scientific research contributed much to new technology.[140]

Hohenberg and Lees argue that, whether it was informal know-how or formal knowledge, information was, with increasing regularity, one of the main inputs of small-scale industry. And large, diversified cities were centers where information accumulated and multiplied. The innovations to which these economies of agglomeration led made these cities pioneers in many new industrial products and processes, which would later be exported to the centers of heavy industry once they had been routinized. "The nature of information as an input to production is that it ceases to be important once a given process becomes routine. At that point other costs—for machines, basic labor, and space—take over, and central cities are at a serious disadvantage. Moreover, economies of scale become critical and . . . very large cities are not especially favored locations for the largest enterprises."[141]

Thus, even though routinization may not be conducive to, and may even preclude, innovation, this loss is offset through the gains derived from economies of scale. Moreover, increasing the command element in the economic mix reduced not only production costs but transaction costs as well. This is indeed how the neoinstitutionalist economist Oliver Williamson explains the replacement of markets by hierarchies. In his view, these two extremes and their hybrids represent different "governance structures" for handling the same transactions. Poor information about a good to be exchanged, opportunist behavior by the partners of exchange, difficulties in drawing sales contracts that foresee all eventualities (as well as other imperfections of real markets) increase the costs of transacting in a decentralized way. At the limit, transaction costs may override the gains from trade and then it may prove profitable to switch from markets to hierarchies as the mode of governing transactions.[142]

Williamson argues, for example, that as any asset develops a high degree of specificity (e.g., one firm buys machinery geared exclusively toward the needs of another firm, or workers develop skills for particular processes), a relationship of dependence develops between the people involved which may leave the door open for opportunist behavior. In this situation, given the much increased costs of defining contracts that counteract the effects of opportunism, it will pay for one company to absorb the other, that is, to replace a relation based on prices by one based on commands. In the case of workers, the transaction costs involved may be those of bargaining over the terms of a contract. The routinization of the production process and the consequent de-skilling of the workers reduces their bargaining power and the consequent costs for managers of transacting in the labor market.[143] However, Williamson's approach, in which an increase in the command element of economic organizations is justified exclusively in terms of efficiency (economizing transaction costs), has been criticized for overlooking the noncontractual benefits (to the managers of firms) of industrial discipline.[144] This is one reason for viewing the development of economic institutions (particularly in the United States) as part of a wider "organizational ecology," which must include military institutions. In the next chapter we will need to widen even more the scope of this "ecology" as we develop Foucault's idea that the efficiency of economic organizations (for example, the factory system) needs to be measured both in terms of economic utility and in terms of political obedience, which is where disciplinary institutions play an important role.

In the nineteenth century there were two more processes benefiting hierarchies over meshworks in the economy. On one hand, as Douglas North argues, as economies complexified (as the amounts of fixed capital increased, for example), the proportion of the gross national product spent on transaction costs also increased. This led to an institutional evolution in which informal constraints were increasingly converted into formal rules and decentralized enforcement replaced by the coercive intervention of central states, in order to keep transaction costs relatively low.[145] On the other hand, the population of commercial organizations inhabiting cities (and the industrial hinterlands these cities animated) underwent dramatic changes. In particular, an organizational form that preexisted the Industrial Revolution but had always been a small part of the population now began to proliferate: the joint-stock company. This type of organization is characterized by a separation of ownership from control: the owners are a dispersed group of stockholders, and control of the company passes from the owner-entrepreneur to the professional manager (or, rather, to a managerial hierarchy).

Galbraith, for example, argues that although joint-stock corporations have boards of directors which represent the owners, in practice this function has become largely ceremonial, particularly in firms where the managers select the members of the board. Ownership is also separated from control by the fact that the managers have a more complete knowledge of the daily operations of the firm. In these circumstances, the strategy of the institution changes from one of maximizing the wealth of the stockholders to one of growth for its own sake, since this increases

the complexity of the operation and hence the need for insider, managerial knowledge.[146]

Interestingly, the most intense proliferation of this organizational form
did not occur in the more industrially advanced British cities but in the
United States. (The British and the Dutch did have joint-stock companies,
particularly the famous Companies of Indias, which were like states within the state.)[147] It was in America that these organizations began a process of enormous growth by swallowing smaller companies, increasingly
replacing markets with hierarchies. Indeed, one economist goes so far as
to say that the reason Britain lost its industrial lead to the United States
by the early twentieth century was precisely because this absorption of
markets by hierarchies did not take place. Britain's problem "was not
that it relied too little, but that it relied too much, on market coordination
of its economic activities."[148] There are many competing explanations
for why large-scale enterprises in which commands increasingly replace
prices as a coordination mechanism failed to develop on British soil, at
least with the same intensity as on the other side of the Atlantic. One
interesting possibility rests on the idea that London (and the rest of England's cities, which fell under its control) was at the time the core of the
Network system (and hence of the now globalized world-economy) and
that, as such, it had the resources of the entire world as its own private
supply zone. (That is, in the nineteenth century, England as a whole may
be seen as a monopoly.) Back in the fourteenth and fifteenth centuries,
when Venice was the core of the European world-economy, "she was far
behind the pioneer cities of Tuscany as regards banking or the formation
of large firms."[149] It is almost, as Braudel suggests, as if the whole of
Venice, whose entire population lent money to the merchants, were a
huge joint-stock company itself, thereby inhibiting the development of
this organizational form within it.

Whatever the reasons for the delay in Britain, the process of separation of ownership from control and the wholesale replacement of markets
with hierarchies were particularly clear in urban settlements in the United
States. This country had witnessed its own acceleration of city building in
the nineteenth century. While the population of towns in 1790 included
only a couple dozen cities, by 1920 there were almost three thousand.[150]
This population included capitals, gateway ports, and industrial towns of
different types, from oppressive antimarket towns like Pittsburgh to more
socially concerned textile mill towns like Lowell, Lawrence, and Manchester.[151] In the later part of the century, this acceleration further intensified
and the percentage of the human population living in urban centers doubled between 1890 and 1920.[152] Industrialization had also intensified, so

that by the turn of the century the United States had become the world's leading manufacturer.

The population of commercial institutions inhabiting American cities underwent an intense wave of internalization of markets by hierarchies. This integration took one of three forms: backward vertical integration, which meant that a manufacturer absorbed its suppliers of raw materials; forward vertical integration, which resulted in the incorporation of a firm's distribution system; and, finally, horizontal integration, which involved taking over other firms in the same industrial specialty.[153] In the second half of the nineteenth century, Chicago's toolmakers and meat packers, Milwaukee's beer producers, New York's textile mills and sewing-machine manufacturers all began a process of forward vertical integration by developing their own nationwide marketing operations, internalizing an economic function previously performed by networks of commissioned salesmen and brokers. While the American economy in 1850 "was one of small businesses with many unintegrated firms dependent upon many marketing middlemen . . . by 1900, contemporary observers were describing a quite different world, a world of vertically integrated big business. A few large firms whose interests spread out over the whole country dominated every major industry."[154] American industrial hierarchies both absorbed their markets and merged among themselves, with the aim of avoiding oligopolistic competition and increasing centralized control:

> The railroads, which were the country's first big business, encouraged other big business in at least two ways in addition to providing the model. . . . They were a cardinal factor in creating a national market, and in doing so, they put a sharper edge on intramural competition. They broke down monopolistic market positions by making it possible for firms to invade each other's territory. To protect themselves from the wounds and bruises of competition, businessmen integrated horizontally as well as vertically, thus giving another boost to big business.[155]

In the northeastern United States, the process of internalization would play an important role in the next great energy intensification: electrification. While independent inventors (such as Edison), who benefited from economies of agglomeration, had developed the first few electrical products, a process of internalization by investors[156] was behind the harnessing of the gravitational energy of Niagara Falls, and it was the latter that transformed electricity from its limited role as a source of illumination to that of a universal form of energy. In the course of this undertaking, crucial technical questions (such as the relative merits of direct versus alternating current) were settled, and the nature of the enterprise itself (a producer of energy, not a supplier of light) was elucidated. The driving force behind the project was a group of bankers who formed the Cataract Construction Company in 1889. They internalized an established company and all its machinery, and set out to face the complex technical and logistical problems of conquering the falls. By 1896, the plant they built was transmitting power to the city of Buffalo, and the electrical utility company as we know it had come into existence.[157]

A product of investor internalization, the electrical industry helped pioneer a new form of absorption: the direct internalization of economies of agglomeration. Unlike its rivals (coal gas for illumination, steam for motorization), electricity was increasingly dependent on formal and informal knowledge for its development. Knowledge, in turn, is an input of production which exacts high transaction costs. Only where patents are perfectly enforceable will information be allowed to flow through markets, else antimarkets will prefer to internalize it into their hierarchies.[158] One way a corporate hierarchy may internalize knowledge is by funding a research laboratory. Although the German organic-chemistry laboratories and Edison's Menlo Park lab were precursors, the first modern industrial laboratory dedicated exclusively to research (as opposed to mere testing) was created by the General Electric Company in the early years of the twentieth century. The G.E. lab, and the many that were later created in its image, may be viewed as an internalized meshwork of skills:

> It is a great strength of the industrial laboratory that it can be both "specialist" and "generalist," permitting an individual to work alone or a team to work together.... The research laboratory provides an individual with access to skills and facilities which greatly increase his capacity. It can at the same time, however, organize a team effort for a specific task and thus create a collective "generalist" with a greater range of skills and knowledge than any individual, no matter how gifted, could possibly acquire in a lifetime.[159]

Although the use of electricity as an energy source owed its origins to urban economies of agglomeration and the information they generate, once those meshworks had been internalized and routinized, electricity's future belonged to economies of scale. Much as the steel industry, which required larger and more sophisticated machinery and plants than iron mills, automatically benefited larger enterprises, so electricity immediately matched the scale at which antimarket institutions operate.[160] The new intensification took place along several fronts. Size, temperature, and pressure were all intensified to generate economies of scale in the *production* process. Voltage, too, was greatly intensified, and positive feedback was created in the *transmission* process as well. Yet, as far as electricity's effect on society, the intensification that mattered most was that of *consumption*, which followed naturally from electricity's multitude of potential uses. In other words, the injection of more and more energy into urban centers would not on its own have produced much of a change, since the uses to which the older forms of energy could be put were limited. At some point urban societies would have reached a point of saturation, and the intensification would have ceased. But electricity simultaneously increased the flow of energy and the potential uses of that energy. Hence, in this case, it was as much the intensity as the *form* of the new energy inputs that mattered.

At the turn of the century, electricity had three possible uses, not to mention a multitude of potential uses (such as computers) that would be realized only later. These three applications were communications, lighting, and mechanical power. The first two were the better known, since electricity had been connected with the flow of information from early on. Stored in batteries, it had powered the telegraph throughout the nineteenth century. Electricity had also powered lighting systems, beginning in the 1870s. But its true transforming power would not depend as much on its role in communications or illumination as in the creation of a new breed of motors that, unlike steam engines, could be *miniaturized*, which permitted a new degree of control over the flow of mechanical energy.[161] The miniaturization of motors allowed the gradual replacement of a centralized engine by a multitude of decentralized ones (even individual tools could now be motorized). Motors began disappearing from view, weaving themselves into the very fabric of reality.

Of course, electricity was not the sole cause of the last great intensification undergone by Western cities. As with earlier intensifications, it was the interplay of several innovations (electricity and electrical products, the automobile and its internal combustion engine, plastics and other synthetic materials, steel and oil) that allowed this intensification to sustain itself.

It is also important to keep in mind that this new web of interlocking technologies did not replace the old one. Although coal lost ground to oil in this century, even as late as the 1960s coal still accounted for half of the world's energy consumption, and its reserves were less depleted that those of oil.[162] Rather than performing a wholesale replacement, the new circuit of triggers and flows inserted itself into the old one. The original loop (coal-iron-steam-cotton), and its newly acquired nodes (railroads, telegraph), continued to function into the twentieth century. The new

91

technologies simply grafted themselves into the previous meshwork, becoming yet other nodes, participating in its self-reproduction and, hence, reproducing themselves. Rather than being left behind, the old circuit simply complexified, losing a few trigger-flow components while gaining many new ones.

Cities began to change under the influence of these new nodes. New York and Chicago in particular experienced an intense electrification and metallization, which resulted in the birth of the skyscraper, an original urban form unique to the United States, prior to World War II. The iron frame, which allowed masonry walls to be replaced with glass, had been pioneered in European cities such as London and Paris. But it was in America that this metallic endoskeleton evolved into the skyscraper. Electric motors in turn allowed elevators to transport people vertically through these huge towers. Chicago pioneered the use of steel and electricity in the construction industry, catalyzed by the great fire of 1871, which destroyed the city's commercial center and literally cleared the way for innovative building techniques to be applied. By the 1890s, Chicago was the world capital of the skyscraper, with New York a close second. But if electricity and steel acted as centripetal forces, making possible the intense human and machine concentrations represented by the new megacities, the internal combustion engine and the automobile had a centrifugal effect, allowing people to move out of central cities into rapidly growing suburban areas. Automobiles, say Hohenberg and Lees, "acted as a solvent rather than a cement to the urban fabric."[163]

The year 1920 marks a turning point in the acceleration of American city building, the moment when the number of Americans living in cities surpassed the number inhabiting rural areas. But 1920 also marks the moment when the growth of central cities was surpassed by the growth at their fringes, the moment urban deconcentration began to intensify. As suburbs started housing more of the urban population than central cities, the latter became part of larger "metropolitan regions" (as they came to be known) and of a new territorial division of labor. Cities lost some of their economic functions to suburbs and industrial hinterlands, and developed specializations in yet other functions (those that were information-intensive). This process was largely unplanned, forming a territorial meshwork of interlocking specializations. As one author puts it, "One might describe the metropolitan region as a giant network of functional relationships in search of a form and a government."[164]

Besides these changes in internal form, the relationship between cities in Europe and in America began to change. In particular, the core of the global Network system shifted in the 1920s from London to New York

City. By the twenties, New York had already enjoyed several decades of financial independence from London. For instance, electrification, unlike New York's earlier intensifications, had not been financed from abroad.[165] A few decades later, after World War I, the United States emerged as a creditor nation, and another maritime metropolis (New York), not a land-locked capital (Washington), would assume the role as core of the global Network system.

However, New York would soon experience a phenomenon whose roots went back several centuries, to the time when nation-states first began to swallow up urban centers: the process of *city killing*. One factor contributing to the depletion of urban autocatalytic dynamics was the unprecedented mobility of large corporations, which, having internalized the benefits of economies of agglomeration, could move headquarters and production facilities with relative ease. Unlike small firms, which are locked in a meshwork of interdependencies with other small enterprises and hence cannot easily move to another city, industrial antimarkets are free to change location between, or outside of, urban centers. And as they move away, large corporations take their internalized meshworks with them, depriving cities of an incalculably valuable resource. Meshworks of small firms may also be destroyed in a more direct way by large organizations using their economies of scale to gain control of markets. In Braudel's words:

> Over the twenty years or so before the crisis of the 1970's, New York—at that time the leading industrial city in the world—saw the decline one after another of the little firms, sometimes employing less than thirty people, which made up its commercial and industrial substance—the huge clothing sector, hundreds of small printers, many food industries and small builders—all contributing to a truly "competitive" world whose little units were both in competition with, yet dependent upon each other. The disorganization of New York was the result of the squeezing out of these thousands of businesses which in the past made it a city where consumers could find in town anything they wanted, produced, stored and sold on the spot. It was the big firms, with the big production units out of town, which ousted the little men.[166]

Antimarket organizations were not the only hierarchical structures engaged in city killing. According to Jacobs, governmental bureaucracies have for centuries been destroying urban meshworks in a variety of ways, a diversity of "transactions of decline" (as she calls them) that result in the loss of positive feedback, or at least in the loss of the special type of economies of agglomeration involved in import-substitution dynamics. Because small cities need a flow of imports to build up the critical mass that results in an explosive episode of replacement dynamics, any government policy that redirects this flow away from them is a potential city killer. Taxing urban centers in order to sustain rural subsidies is one example, as is the promotion of trade between large and small cities, since a large city will attempt to transform a smaller city into a supply zone. (As we observed earlier, volatile trade requires backward cities to use each other symbiotically.)[167]

To return to our main argument, despite the loss of vitality of many cities, the great autocatalytic loop of triggers and flows continued to complexify by acquiring new nodes (electricity, automobiles), which allowed it to circumvent internal limits to its growth (such as a saturation of the urban demand for more and more energy). The continuing growth also depended, of course, on other factors, such as the availability of relatively cheap energy sources, and this in turn depended on the ability of Western nations to transform the rest of the world into a vast periphery, or supply zone. We will return to the question of colonialism in the next chapter, but for now it will suffice to note that, unlike the original circuit of triggers and flows in Britain during the Industrial Revolution, the resource nodes in the expanded version of the loop (the second industrial revolution) had long been international. (Western cities became painfully aware of their long dependence on underpriced energy — and hence their dependence on their global supply zones — during the oil crisis of the 1970s.) The autocatalytic loop became increasingly dependent, too, on the flow of information. And this flow, in turn, began to be affected by the creation of new institutions: the research laboratory and the technical university. As Peter Drucker writes:

> Few of the major figures in 19th century technology received much formal education. The typical inventor was a mechanic who began his apprenticeship at age fourteen or earlier. The few who had gone to college [Eli Whitney, Samuel Morse] had not, as a rule, been trained in technology or science, but were liberal arts students.... Technological invention and the development of industries based on new knowledge were in the hands of craftsmen and artisans with little scientific education but a great deal of mechanical genius.... The 19th century was also the era of technical-university building. Of the major technical institutions only one, the Ecole Polytechnique in Paris, antedates the century.... But by 1901, when the California Institute of Technology in Pasadena admitted its first class, virtually every one of the major technical colleges active in the Western world today had already

come into being. Still, in the opening decades of the 20th century the momentum of technological progress was being carried by the self-taught mechanic without specific technical or scientific education.[168]

The switch from the self-taught inventor of the nineteenth century to the industrial laboratory of the twentieth, with its staff of technical-university graduates, involved a reversal in the balance of power between informal and formal knowledge. Still later on, the advent of computers (which are basically automated formal systems) appeared to consolidate the victory of analytical over embodied knowledge, to the point where the difference itself seemed to vanish for all but a few philosophers.[169]

According to Galbraith, the enlarged role that knowledge began to play as an input to production processes (as well as in other areas of corporate activity, such as marketing) had a significant impact on the governance structure of large economic organizations, acting as a counterbalance to the increased amount of command elements in their mix. Despite the existence of managerial hierarchies in most corporations, the decision-making processes within these institutions are not based entirely on rank and formal authority, but on committees, an apparatus for group decision making (which he calls the "technostructure"). These committees serve as a means for pooling knowledge, formal and informal, and as mechanisms for testing the relevance of collective opinions. Top management tends simply to ratify the decisions made by these collective bodies, particularly in situations where the decisions to be made are not routine.[170]

The intensification of the flow of knowledge also affected the dynamics of cities and their industrial hinterlands. A recent study of two industrial hinterlands — "Silicon Valley" in Northern California and Route 128 near Boston, both of which developed in close contact with technical universities (Stanford and Massachusetts Institute of Technology, respectively) — illustrates the effects of this intensification. The study observes that:

> Silicon Valley has a decentralized industrial system that is organized around regional networks. Like firms in Japan, and parts of Germany and Italy, Silicon Valley companies tend to draw on local knowledge and relationships to create new markets, products, and applications. These specialist firms compete intensely while at the same time learning from one another about changing markets and technologies. The region's dense social networks and open labor markets encourage experimentation and entrepreneurship. The boundaries within firms are porous, as are those between firms themselves and between firms and local institutions such as trade associations and universities.[171]

The growth of this region owed very little to large financial flows from governmental and military institutions. Silicon Valley did not develop so much by economies of scale as by the benefits derived from an agglomeration of visionary engineers, specialized consultants, and financial entrepreneurs. Engineers moved often from one firm to another, developing loyalties to the craft and the region's networks, not to any particular corporation. This continual migration, together with an unusual (for corporations) culture of information sharing among the local producers, ensured that new formal and informal knowledge would diffuse rapidly through the region. Business associations fostered collaboration between small and medium-sized companies. Risk taking and innovation were preferred over stability and routinization. (Of course, this does not mean that there were not large, routinized firms in Silicon Valley, only that they did not dominate the mix.) Not so on Route 128:

> While Silicon Valley producers of the 1970's were embedded in, and inseparable from, intricate social and technical networks, the Route 128 region came to be dominated by a small number of highly self-sufficient corporations. Consonant with New England's two century old manufacturing tradition, Route 128 firms sought to preserve their independence by internalizing a wide range of activities. As a result, secrecy and corporate loyalty govern relations between firms and their customers, suppliers, and competitors, reinforcing a regional culture of stability and self-reliance. Corporate hierarchies ensured that authority remains centralized and information flows vertically. The boundaries between and within firms and between firms and local institutions thus remain far more distinct.[172]

Before the recession of the 1980s, both Silicon Valley and Route 128 had been continuously expanding, one on economies of agglomeration and the other on economies of scale (or, rather, mixtures dominated by one type or the other); nonetheless, they both felt the full impact of the downturn. In response to hard times, some large Silicon Valley firms, ignoring the dynamics behind the region's success, began gearing production toward economies of scale, transferring the manufacture of certain parts to other regions and internalizing activities previously performed by smaller firms. Yet the intensification of routinization and internalization in Silicon Valley was not a constitutive part of the region (as it was on Route 128), which meant that the old meshwork system could be revived. And that is, in fact, what happened. Silicon Valley's regional networks were reenergized, through the birth of new firms in the old pattern, and the region has now returned to its former dynamic state, unlike the command-heavy Route 128, which continues to stagnate. This shows that, while economies of scale and economies of agglomeration, as forms of positive feedback, both promote growth, only the latter endows firms with the resilience and adaptability needed to cope with adverse economic conditions.

The case of Silicon Valley and Route 128 shows that there are several viable lines of development for future production systems, much as there were alternative forms of industrialization in previous centuries. Paradoxically, the computerized products manufactured in these two industrial hinterlands, and the further intensification in the flow of knowledge that computers allow, could push the evolution of industrial production in either direction, to increase or decrease the relative proportions of command and self-organization.

On one hand, computers may become the machines that finally eliminate human beings and their flexible skills from industrial production, as in fully automated factories. Maturana notes that one characteristic of autocatalytic loops is that their internal states determine most of their behavior, with external stimuli playing the role of triggers. He compares this to push-button machines whose behavior is not caused by the pushing of a button, only triggered by it.[173] Automated factories are very complex push-button machines of this type and, as such, *planned autocatalytic loops*. Indeed, as late as the 1960s, a routinized, rationalized production process that generated economies of scale was thought by many to be the perfect example of a whole that is more than the sum of its parts. That so-called systems approach celebrated routinization as the crowning achievement of modern science.[174] Today we know that planned loops of triggers and flows are only one of a number of systems that exhibit emergent properties, and that spontaneously generated loops may be more adaptive and resilient than rigidly planned ones.[175]

Automation results in self-sustaining autocatalytic loops of routines, with a limited capacity for spontaneous growth. These loops emerge and grow by corporate planning, so they can be only as good as the planners themselves. On the other hand, instead of aiding the growth of self-sufficient corporations, computers can be used to create a network out of a collection of small firms, as happened in some industrial hinterlands in Europe, allowing economies of agglomeration to compensate for the lack of scale of the individual firms.[176] In this case, the abilities of the individuals involved will be amplified by processes of self-organization occurring at the institutional and regional levels. By facilitating the formation of meshworks of complementary economic functions, the computers created in industrial hinterlands could allow urban centers to recover the rich

nonlinear dynamics of earlier production methods, such as import-substitution dynamics.

If something like this were to happen, these regions would simply be
repaying a very old debt to cities. Industrial hinterlands have always
emerged in close connection with dynamic urban centers, spawned and
nourished by cities and towns enjoying some kind of positive feedback
from their agglomeration of skills and economic functions. Cities that
served mostly as administrative centers, with more command than market components, did not animate active industrial regions beyond their
borders. London, Amsterdam, Paris, Los Angeles, New York, São Paulo,
Singapore, and Seoul did, while Madrid, Lisbon, Atlanta, Buenos Aires,
Manila, and Canton did not. According to Jacobs, the latter lacked the
volatility in trade and the dynamism of small-producer networks needed
to infuse life into a city's regions, as opposed to merely exploiting them
as resource depots.[177]

Needless to say, computers will not magically produce a quick technological fix to urban problems. For one thing, they may still evolve in the
direction of routinization, further eroding the combinatorial richness of
knowledge and making flows of information ever more sterile. The digital
revolution should be thought of as one more element added to a complex
mix, fully coexisting with older components (energetic and material), not
all of which have been left in the past. In other words, digital machinery
is simply a new node that has been grafted on the expanding autocatalytic loop. Far from having brought society to a new stage of its development, the information stage, computers have simply intensified the flow
of knowledge, a flow which, like any other catalyst, still needs matter and
energy flows to be effective.

There is one final institutional development that needs to be mentioned here: the transnational corporation. Although government and military institutions evolved side by side with big business, forming a true
meshwork of hierarchies, a recent intensification of the mobility that has
always characterized antimarkets has allowed them to transcend national
boundaries and hence their interlocking relationships with the state.
(Transnational corporations are not a new phenomenon, but they used to
form a small fraction of the total population of urban firms.) The routinization of production and the internalization of markets are now carried
on at a global level, while powerful computers allow the centralized control of geographically dispersed activities. According to some analysts,
the internationalization of antimarket institutions (or at least the intensification of this process) was indeed brought about by advances in the science of centralization (for example, in operations research, which was

developed by the military during World War II) and by the use of large computers to coordinate and monitor compliance with central plans.[178]

In this way, many corporations have now become truly independent of any particular country, much as decades ago they became independent of cities. Indeed, nation-states have become obstacles for the expansion of antimarket institutions, since the achievement of economies of scale at an international level demands the destruction of the regulations with which independent countries attempt to control the flows of money, goods, and information across national borders.

Despite the fact that meshwork-generating processes are active today in several parts of the globe, hierarchical structures enjoy a commanding, two- or three-hundred-year lead, which could very well decide the issue, particularly now that processes of homogenization have become international. But even if the future turns out to belong to hierarchies, this will not occur because a "law of capitalism" somehow determined the outcome from above. Human history is a narrative of contingencies, not necessities, of missed opportunities to follow different routes of development, not of a unilinear succession of ways to convert energy, matter, and information into cultural products. If command structures end up prevailing over self-organized ones, this itself will be a contingent historical fact in need of explanation in concrete historical terms. I have already suggested here that a multiplicity of institutions (economic, political, and military) will enter into this explanation. A more detailed analysis of the process through which homogenizing forces came to overwhelm those promoting heterogenization will in fact involve a wider variety of organizations (including schools, hospitals, and prisons).

In the next chapter, we will explore other aspects of the accumulation of hierarchical structures within the European and American exoskeleton. Examining the role that these institutions played will allow us to put some flesh on the bare bones of our account of Western institutional and urban history.

2: FLESH

AND GENES

Biological History: 1000–1700 A.D.

In the eyes of many human beings, life appears to be a unique and special phenomenon. There is, of course, some truth to this belief, since no other planet is known to bear a rich and complex biosphere. However, this view betrays an "organic chauvinism" that leads us to underestimate the vitality of the processes of self-organization in other spheres

of reality. It can also make us forget that,
despite the many differences between them,
living creatures and their inorganic counterparts share a crucial dependence on intense
flows of energy and materials. In many
respects the circulation is what matters, not
the particular forms that it causes to emerge.
As the biogeographer Ian G. Simmons puts
it, "The flows of energy and mineral nutrients through an ecosystem manifest themselves as actual animals and plants of a particular species."[1] Our organic bodies are, in
this sense, nothing but temporary coagulations in these flows: we capture in our bodies
a certain portion of the flow at birth, then
release it again when we die and microorganisms transform us into a new batch of
raw materials.

The main form of matter-energy flow in
the biosphere is the circulation of flesh in
food chains. Flesh, or "biomass," circulates
continuously from plants to herbivores,
and from herbivores to carnivores, giving the
ecosystem its stability and resilience. (This
basic food chain is in reality only one among
several, forming a system of interlocking
chains referred to as a "food web.") The foun-

dation of any food web is its plants, which "bite" into the stream of solar radiation, capturing some of it as sugars by means of photosynthesis. Plants are the only nonparasitic creatures in an ecosystem, its primary producers, while the animals who eat flesh (plant or animal) are mere consumers. The complex microflora and microfauna that process the ecosystem's waste are as important as plants, since these organisms remineralize and reinject dead plant and animal bodies back into the web.[2] Compared to plants and microorganisms, "higher" animals are just fancy decorations in an ecosystem, consuming and transforming biomass with decreasing efficiency as their size increases.[3]

For this reason, the emergence of an ecosystem is typically described as a *succession of plant assemblages* that interact with each other, passing through several stable states until they reach a "climax." A temperate forest, of the type that characterizes the European continent, for example, begins as an assemblage of lichen and moss, followed by scrubby birch and aspen, then pine forest, and finally a mature oak, lime, elm, and beech forest.[4] Although it may appear otherwise, this process of succession does not have the climax state as its goal. Rather, the emergence of an ecosystem is a blind groping from stable state to stable state in which each plant assemblage creates the conditions that stabilize the next one. A variety of historical constraints (energetic, material, dynamical) determine at some point that there is no other stable state attainable from the current one, and so the process climaxes. This is, of course, just another example of a meshwork of heterogeneous elements evolving by drift. A more realistic model of this meshwork would have to include microorganisms, the myriad insects and other small animals that play key roles in the flow of biomass, and even some "decorative" large predators, like tigers, wolves, or early humans.

This section explores the relationships between medieval cities and towns and the ecosystem in which they grew—not only the forests they devoured as they proliferated but also all the other interactions they maintained with biological entities, especially microorganisms. Here we will argue that even though plants were in a way submitted to the control of the towns, microbes resisted control much longer (if indeed we can say that antibiotics have finally brought them under our command, which may not be quite true). And then, of course, we must consider that other uncontrollable element of ecosystems, the climate. Both infectious diseases and changing weather patterns played a great role in urban history, making epidemics and famines part of the "biological regime" that dominated urban and rural life until the eighteenth century.

From a different perspective, cities and towns may themselves be considered ecosystems, at least to the extent that biomass circulates through them to feed their inhabitants. The diagram of this circulation, however, must include processes occurring outside cities and towns because urban centers have always depended on their countrysides for food. In human-made ecosystems, the inhabitants of the surrounding villages are the primary producers while city dwellers, despite their cultural sophistication, are mere consumers. Moreover, this parasitic relationship can be reproduced at a larger scale. In the early sixteenth century, for example, as cities grew and developed trade links with one another, their food began to flow from ever remoter supply zones. First eastern Europe was transformed into a vast "countryside" for the urban complex to its west, then America and other foreign lands were converted into resource depots to feed western European cities.

Thus, ours will be a dual story, one tracing our biological connections to nonhuman life, the other describing the gradual conversion of the world into a supply region to fuel European urban growth. We begin by discussing the principal difference between natural and urban ecosystems: their degree of homogeneity and heterogeneity.

Ecologists have learned from their empirical study of ecosystems that there is a close relationship between stability and the degree of species heterogeneity in a food web. However, the nature of the connection between the two is not yet fully understood. In the early seventies, some mathematical models of ecosystems suggested that there may not even be a connection: webs of *randomly assembled* species tended to become more unstable as new species were added; diversity bred instability. However, all that those models proved was that real ecosystems are *not* random assemblages of species, but self-organized meshworks in which species are interconnected by their functional complementarities: prey and predator, host and parasite.[5] According to one ecologist, heterogeneity endows these meshworks not so much with stability (the capacity to maintain a state with relatively minor internal fluctuations) as with resilience (the capacity to absorb major external and internal fluctuations by switching between several alternative stable states).[6] Continental forests are an example of these resilient webs of interlocked species. Islands far from the mainland, on the other hand, are more homogeneous and less capable of absorbing shocks and may be drastically destabilized by a sudden influx of a new species.

The cities that began multiplying in Europe at the beginning of the millennium were like so many islands in the middle of a large temperate forest in its climax state, dominated by oaks and elms. Cities are like islands in two different ways. In terms of climate, cities are "heat islands," separated from their countrysides by a sharp difference in temperature.[7] Large furnaces and machines that emit heat, a mineral infrastructure that stores heat from the sun and then releases it at night, and low evapotranspiration are among the factors that contribute to making large cities concentrations of waste energy. In medieval times, of course, only a few regional capitals and gateway ports (if any) had mineralized and industrialized enough to become heat islands. But all medieval towns big and small were islands in another respect: their low degree of species heterogeneity. A typical medieval town can be described as a tightly packed assemblage of humans, a few species of animals and plants, and, as one writer has put it, "a lumpen-proletariat of insects."[8]

Because towns are necessarily parasitic on their rural surroundings, urban ecosystems encompass more than what is found inside their walls. A town with three thousand inhabitants, a medium-sized town in the Middle Ages, needed to control the lands of at least ten villages around it (an area of approximately five square miles) to ensure a constant supply of edible biomass. Thus, although density of population is the criterion normally used to define an urban center, Fernand Braudel argues that the division of labor between food producers and consumers (and the power needed to impose and maintain it) is the true defining trait of urban life.[9] We should not imagine, however, that the medieval distinction between the urban and the rural was as sharp as it is today. "Even the large towns continued to engage in rural activities up to the eighteenth century. In the West they therefore housed shepherds, gamekeepers, agricultural workers and vinegrowers (even in Paris). Every town generally owned a surrounding area of gardens and orchards inside and outside its walls.... In the middle ages the noise of the flail could be heard right up to the Rathaus in Ulm, Augsburg and Nuremburg. Pigs were reared in freedom in the streets."[10]

The main characteristic of an urban ecosystem is its homogeneity: human beings *shorten all food chains* in the web, eliminate most intermediaries and focus all biomass flows on themselves.[11] Whenever an outside species tries to insert itself into one of these chains, to start the process of complexification again, it is ruthlessly expunged as a "weed" (a term that includes "animal weeds" such as rats and mice). Medieval towns were, in this respect, no exception. Moreover, the agricultural lands that fed these towns were themselves simplifications of the forests they had replaced. When a piece of forest was cleared to create arable land, an assemblage of plants in its climax state was driven back to its very first state of succession, its species composition homogenized and its energy and nutrients redirected toward a single center. (Yet, for the same reason, it was transformed into a place where plant species with "opportunistic" reproductive strategies [i.e., weeds] could multiply.)

The same held true with respect to animals. Several domesticated species (pigs, cattle, goats) may be considered *biomass converters*, which aid the process of shortening and redirecting food chains. For example, cattle and goats transform indigestible biomass (leaves, grass, sprouts) into edible flesh and milk. Pigs are even more efficient converters (one-fifth of the carbohydrates they eat are transformed into protein), but they feed mostly on sources that are also suitable for human consumption.[12] They can nevertheless serve as living storage devices for unpredicted surpluses. Together, humans and their "extended family" of domesticates, as the historian Alfred Crosby calls it, transformed a heterogeneous meshwork of species (a temperate forest) into a homogeneous hierarchy, since all biomass now flowed toward a single point at the top. In a sense, a complex food web was replaced by a simplified food pyramid, at least in those areas where urbanization had triumphed.

This homogenization, however, had to be maintained through the sheer weight of human numbers. Whenever the human population declined, the animals and plants that were excluded from the urban ecosystem made a comeback. Roughly speaking, Europe's population increased between 1100 and 1350 and between 1450 and 1650; it declined between 1350 and 1450 and again between 1650 and 1750. In the periods of decline, humans had to struggle to keep their place at the top of the pyramid:

> The whole of Europe, from the Urals to the Straits of Gibraltar, was the domain of wolves, and bears roamed in all its mountains. The omnipresence of wolves and the attention they aroused make wolf-hunting an index of the health of the countryside, and even of the towns, and of the character of the year gone by. A momentary inattention, an economic setback, a rough winter, and they multiplied. In 1420 packs entered Paris through a breach in the ramparts or unguarded gates. They were there again in September 1438, attacking people this time outside the town, between Montmartre and the Saint-Antoine gate.[13]

Large predators continued their visitations until the end of the eighteenth century, by which time human hunters had nearly driven them to extinction. And yet they were not the only species for whom human beings were a food source. Of greater importance, and of more enduring influence, were the "micropredators," the diseases that ate human flesh from within. Contagious diseases and their hosts form complex, nonlinear dynamical systems with several possible states. When the population of hosts is insufficient, or insufficiently packed, making contagion difficult for the microorganism, the dynamical system enters an unstable state called "epidemic," and the population of germs grows explosively until it burns out its human fuel. When, on the contrary, overall population and population density are beyond a critical threshold, so that there is always a fresh supply of flesh for the parasites to infect (typically small children), after a few epidemics the dynamical system stabilizes into what is called an "endemic" state. Human survivors of the disease become immune, the microorganisms lose some of their virulence and microbe and host enter into a state of mutual accommodation. In William McNeill's words:

> Only in communities of several thousand persons, where encounters with others attain sufficient frequency to allow infection to spread unceasingly from one individual to another, can such diseases persist. These communities are what we call civilized: large, complexly organized, densely populated, and without exception directed and dominated by cities. Infectious

bacterial and viral diseases that pass directly from human to human with no intermediate host are therefore the diseases of civilization par excellence: the peculiar hallmark and burden of cities and of countryside in contact with cities. They are familiar to almost all contemporary humankind as the ordinary diseases of childhood: measles, mumps, whooping cough, smallpox and the rest.... Most and probably all of the distinctive infectious diseases of civilization transferred to human populations from animal herds. Contacts were closer with the domesticated species, so it is not surprising to find that many of our common infectious diseases have recognizable affinities with one or another disease afflicting domesticated animals. Measles, for example, is probably related to rinderpest and/or canine distemper; smallpox is certainly connected closely with cowpox ... influenza is shared by humans and hogs.[14]

Medieval cities, with their intimate packing of domesticated animals and humans, were veritable "epidemiological laboratories." They offered certain microorganisms the perfect habitat in which to evolve novel variants. Since their very existence would go unrecognized for many centuries, this crucial component of urban ecosystems was effectively outside of human control. Although quarantine measures existed in Europe since the fifteenth century, most cultural accommodations to infectious disease were habits and routines that developed without a conscious plan, by trial and error. These were, in a sense, cultural materials that accumulated unconsciously, sorted out by the pressure of the parasites themselves. Hence, germs and humans formed a meshwork, coevolving through drift, in stark contrast with the rest of the food hierarchy at the service of urban culture.

It is easy to discount the importance of energy and nutrient flows by unduly emphasizing the cultural elements that inevitably flow alongside them. For example, Claude Lévi-Strauss pointed out decades ago that biomass does not enter human society in its "natural" state: it is at the very least processed through the "civilizing" power of fire. In turn, the difference between raw and cooked biomass becomes a largely symbolic opposition, appropriated by myth and legend.[15] Culture also regulates the flow of flesh, distinguishing between taboo, sacred, and everyday foods. The increasing elaboration of sauces and complex dishes which began in Europe in the fifteenth century (and in China and Islam much earlier) added more and more layers of culture to the circulation of raw matter-energy. However, these cultural additives, important as they were, should not blind us to the fact that ultimately it was still the *nutritional value* of the flow that mattered. Nothing serves better to remind us of this fact than the recurrent famines that plagued Europe and other continents, not only in medieval times but until the very eve of the Industrial Revolution. In extreme cases, people would not only eat biomass that had not been culturally sanctioned (such as grass, bark, or even soil), but, more importantly, they would break the most powerful of alimentary taboos and eat human flesh.

> Famine recurred so insistently for centuries on end that it became incorporated into man's biological regime and built into his daily life. Dearth and penury were continual and familiar even in Europe, despite its privileged position. A few overfed rich do not alter the rule. It could not have been otherwise. Cereal yields were poor; two consecutive bad harvests spelled disaster.... For these and other reasons famine only disappeared from the West at the close of the eighteenth century, or even later.... A privileged country like France is said to have experienced 10 general famines during the tenth century; 26 in the eleventh; 2 in the twelfth; 4 in the fourteenth; 7 in the fifteenth; 13 in the sixteenth; 11 in the seventeenth and 16 in the eighteenth. We obviously offered this eighteenth century summary without guarantee as to its accuracy: the only risk it runs is of over-optimism, because it omits the hundreds and hundreds of local famines.[16]

Famines and epidemics were two biological phenomena that competed in importance with the purely cultural phenomena of the times. Culture is not a completely separate sphere of reality, but instead mixes and blends with flows of organic (and even mineral) materials. So far we have emphasized only one of these organic flows—biomass—but of equal importance is the flow of genetic materials through generations. Without this flow, organized flesh would exist in forms as ephemeral as hurricanes (and other nonorganic self-organized entities), and, moreover, it could not evolve. Since evolutionary processes far exceed the life span of individuals, any significant accumulation of adaptive traits requires genetic materials to be registered and stored.

In the view which dominated the West for two millennia the traits that define a given species were necessarily tied together for all time since they were expressions of an eternal essence. Today we know that there is nothing necessary about these accumulations. Species are historical constructions, their defining traits a purely contingent collection assembled by means of selection pressures, which act as a genetic sorting process. In a very real sense, much as our bodies are temporary coagulations in the flow of biomass, they are also passing constructions in the flow of genetic materials. As Richard Dawkins has put it, plants and animals are

merely "survival machines" that have been built to house and perpetuate
the flow of genes, or replicators:

> Replicators began not merely to exist, but to construct for themselves con-
> tainers, vehicles for their continued existence. The replicators that survived
> were the ones that built survival machines for themselves to live in....
> Now they swarm in huge colonies, safe inside gigantic lumbering robots,
> sealed off from the outside world, communicating with it by tortuous, indi-
> rect routes, manipulating it by remote control.[17]

For the biogeographer, the flow of biomass through food webs is
paramount; for the evolutionary biologist, the flow of genes through generations is what matters chiefly. It is clear, however, that the bodies of
animals and plants are transient agglomerations of materials derived
from both of these flows, and not only for the obvious reason that living
creatures must eat (and avoid being eaten) to successfully reproduce.
A more fundamental reason is that the very structural and functional
properties of these bodies cannot be explained in terms of genetic materials alone. Between the information coded into genes and the adaptive
traits of a plant or animal (i.e., between genotype and phenotype),
there are several layers of self-organizing processes, each sustained by
endogenously generated stable states, themselves the product of matter-
energy flow. Genes are not a blueprint for the generation of organic
structure and function, an idea implying that genetic materials predefine
a form that is imposed on a passive flesh. Rather, genes and their products act as constraints on a variety of processes that spontaneously
generate order, in a way *teasing out* a form from active (and morpho-
genetically pregnant) flesh.[18]

Unlike an ecosystem, which is a meshwork of highly heterogeneous
elements, the gene pool of a species may be seen as a hierarchy of
homogeneous elements. As the physicist Howard Pattee has argued, the
crucial function of genes is to force individual molecules within a cell to
obey the cell itself, and similarly for individual cells in a tissue, individual
tissues in an organ, and individual organs in an organism. At each rank
of the hierarchy, the genes' purpose is to constrain the lower level to
behave in ways determined by the immediately upper level.[19] If we imagine a case in which the selection pressures on a species have had the
time and opportunity to work themselves out (i.e., to eliminate many
genes from the pool and drive others to fixation), the resultant species
will indeed be a very homogeneous entity.[20] Of course, in reality most
species retain a degree of heterogeneity, particularly if the selecting envi-

ronment is itself heterogeneous in time or space. Besides, a totally homogeneous species would be incapable of evolving, since natural selection requires variation in the gene pool as its raw material. Nevertheless, compared to ecosystems, the gene pool of a species may be seen as a structure with more command elements in its mix.

Although highly homogeneous, the gene pool of the human species is still variable due to the large variety of ecosystems that humans have colonized, as well as to cultural taboos against interracial marriage. However, whatever heterogeneity remains in the human gene pool affects only our outward appearance and has little adaptive value, with some exceptions. For example, in northern Europe during the Middle Ages, there was a gene coding for an enzyme that allowed adult humans to digest raw milk. Elsewhere, in the populations of China and Islam, for instance, the gene did not exist, so milk had to be culturally processed (transformed into cheese or yogurt) before it could be digested. Another gene, which was distributed to some degree along the Mediterranean but was much more prevalent on the west coast of Africa, allowed its human carriers to resist "being digested" by malarial parasites.[21]

Most human traits are not, of course, determined by a single gene. Skin color, for instance, involves several genes (or more technically, pairs of alleles, alternative genes for the same position in a chromosome). More importantly, most of the genes that aren't common to all human communities define literally superficial traits: skin color, hair form, body shape, and stature. Despite the fact that these traits may have some adaptive significance, the real importance of this heterogeneous "outer shell" is our use of it as a basis for cultural differentiation and racial stereotyping. Truly objective analysis (objective, that is, in comparison to the caricatures of objectivity that Social Darwinists and eugenicists have given us) of the genetic makeup of the body as a whole reveals a stark genetic homogeneity. Interestingly, the genetic variation among individuals of a particular race is greater than the variation between races:

> Of all genetic variation, 85% is between individual people within a nation or tribe.... The remaining variation is split evenly between variation between nations within a race and variation between one major race and another. To put the matter crudely, if after a great cataclysm, only Africans were left alive, the human species would have retained 93% of its total genetic variation, although the species as a whole would be darker skinned. If the cataclysm were even more extreme and only the Xhosa people of the southern tip of Africa survived, the human species would still retain 80% of its genetic variation.[22]

113

The genes that define the "outer shell" (as well as those few that involve biologically important functions, such as resistance to malaria or the ability to digest raw milk) evolved in historical times, which proves that the human gene pool is still changing. But this kind of change ("geologically" slow change) has not played the central role in the dynamics of the human gene pool. That honor is reserved for large migratory movements that mixed hitherto separate populations. For example, the medieval distribution of blood types owed more to ancient migrations than to natural or cultural selection.[23] From the genetic perspective, the causes of human migration (a famine, for instance) are less important than its effects: the homogenizing or heterogenizing consequences of injecting DNA from one local gene pool into another. "Migration is of the greatest genetic relevance. It is the vehicle for the mechanism of evolution that today is producing the greatest evolutionary effect, allowing the incorporation of new genes into established gene pools, enhancing intrapopulation and reducing interpopulation variability."[24]

When human migration is not a movement into previously empty space, it involves the invasion of other people's lands. In terms of its effects on the local gene pool, we may distinguish those cases involving the extermination of the local population (and hence a replacement of one gene pool by another) from those where the aim is to subjugate the locals and use them as a workforce. In this second case there is coexistence between groups, which allows a small trickle of genes to pass between the two groups, despite the social barriers separating one pool from another. This genetic exchange typically occurs from the conqueror's to the conquered's pool.[25]

Several invasions played important roles in shaping the composition of European gene pools. Luigi Cavalli Sforza has discovered in the distribution of genetic materials in present-day Europe an almost circular pattern of some of its components, with its center in the Middle East. After ruling out the hypothesis that selection pressures could have generated this circular gradient (there does not seem to have been enough time for this to happen spontaneously), he has concluded that it was produced by an ancient invasion, which brought agriculture from its place of origin in the Fertile Crescent to the European continent then inhabited by populations of hunter-gatherers. The long- and widely held belief that agriculture was intrinsically superior to hunting and gathering, and hence that it had spread by the *diffusion of ideas*, has been largely refuted by recent research.[26] The old way of obtaining food was as efficient as the new one, so agriculture could not have won over the European population because of its intrinsic superiority; instead, invasion and replacement of some local populations played a key role in spreading the new economic system across Europe. Sforza's computer simulations, however, indicate that to generate the circular pattern we need to allow some acculturation of the remaining hunter-gatherers, involving both cross-marriage and adoption of the new technology.

Although some aspects of culture, the least normative and binding, do travel freely from mind to mind (and from culture to culture), other aspects more central to a society seem to migrate alongside its genes. According to Sforza, languages are a good example of cultural materials that are spread through invasions. Linguistic norms do not diffuse easily from culture to culture (with the exception of individual words), so local languages are easier to kill by eliminating their speakers than to change by local adoption of foreign norms. Another portion of medieval Europe's gene pool was contributed by Indo-European invaders who brought genetic and linguistic materials to the continent and exterminated many local communities and languages.

Medieval European gene pools were also affected by the coexistence of (and gene flow between) different pools. (The expansion and retreat of the Roman Empire and the gene flow between Latin and Germanic pools belong to this category, as do the genes that arrived with the Mongol and Moor invasions, and those spread by the Jewish Diaspora.)[27] The intensity and form of this gene flow were, in turn, affected by cultural institutions: the degree to which marriage occurred outside the group (the degree of exogamy) or the distribution of marriage distances (longer for urban than for rural marriages), for example.[28] In consequence of the various patterns of migration through Europe over the millennia, the entities we designate as "races" today are simply the historical outcomes of these homogenizing genetic flows, and racial groups are differentiated only by their "outer shell":

> Britons, so conscious of their race, are, in fact, an amalgam of the Beaker Folk of the Bronze Age, the Indo-European Celts of the first millennium B.C., the Angles, Saxons, Jutes, and Picts of the first millennium A.D., and finally the Vikings and their parvenu grandchildren, the Normans.... [Hence] the notion that there are stable, pure races that only now are in danger of mixing under the influence of modern industrial culture is nonsense. There may indeed be endogamous groups, largely biologically isolated by geography and culture from their neighbors, such as the Pygmies of the Ituri Forest, but these are rare and not perfectly isolated in any event.[29]

Another crucial role migration plays in urban dynamics affects less the composition of a city's gene pool than the vital processes of the city themselves. Medieval towns, and indeed all cities up to the late nineteenth century, were not self-reproducing entities. That is, they did not reproduce their population by simply combining the flow of biomass from the countryside with the genes that had accumulated within their walls. Death rates in urban centers exceeded birthrates for many centuries (mortality rates among infants and the poor were especially high), so cities were always in need of migrants from the countryside. Sixteenth-century London, for example, needed about five thousand rural migrants a year.[30] And, of course, since many of these immigrants were poor, their mortality rates (and even more so, their children's) increased the moment they passed through the city gates, which explains why towns were commonly referred to as "death traps." "Yet towns, particularly smaller central places (as opposed to ports, proto-industrial cities, or great capitals), were by no means always death traps.... Infant mortality, the key component in normal times [has been calculated to be] equal for rural areas and smaller market towns: 25 to 33 percent of the children up to five years, as opposed to 40 percent to 50 percent in larger cities."[31]

In the nineteenth century, improved water treatment (and other sanitation policies) and mutual adaptation between humans and microorganisms began to reverse the trend and urban birthrates climbed above death rates. But before that (and in many places, a long time afterward) towns were as dependent on rural areas for the influx of genes as they were for the influx of food. Genetic materials from rural gene pools did not, of course, mix freely with those of the city's own gene pool (i.e., the genes of legitimate citizens of the city, who could transmit their rights and obligations to their progeny). Rather, the two pools coexisted and exchanged small flows of genes. For instance, a typical way of gaining citizenship was to marry a citizen's daughter (hence injecting outside genes); and, of course, citizens' genes found their way illegitimately into the migrant population's pool.

This brings us to the question of the social structure of urban centers. So far we have described urban ecosystems as pyramids in which shortened food chains redirect all energy toward the apex, but the existence of social classes implies that the apex itself has a hierarchical structure; that is, it is divided into several *niches* arranged in ranks. *Niche* is the term used by ecologists to define the position of a given species in a food web. It takes into consideration the energy used in searching out and obtaining food, as well as that spent in avoiding being eaten. Each species has its own peculiar way of performing these two tasks, and these behavioral and physiological adaptations define its "job," or niche, in an ecosystem. The ecologist Paul Colinvaux has proposed that, to the extent that different social classes do not have equal access to different types of food (and other energy resources), they might be said to be *social niches*.[32] In the Middle Ages, for instance, many peasants survived on a monotonous diet of bread, gruel, roots, and cooked tubers. They had, in Colinvaux's terminology, a very narrow niche. The elites, on the other hand, whether feudal or urban, had access to a larger variety of food-stuffs, including large quantities of meat and luxury items (e.g., spices). They had a wide niche. In reality, of course, things were more complex and changed over time.

Colinvaux's general point, however, seems to apply regardless of chang-ing historical details. He argues that, just as wild animals must adjust the timing and quantity of their reproductive output (e.g., breeding season and clutch size) to square with the resources available to them, so, too, must humans. In particular, he argues that there is a close relationship between niche width and number of offspring. Peasants and the urban poor, particularly recent immigrants, lived in a penurious but inexpensive narrow niche, so their reproductive "calculations" led them to conclude that they could afford many children. Wealthier classes, on the other hand, desirous of raising wide-niche children, "calculated" that they could afford fewer progeny.[33]

This line of argument corresponds with the populational phenomenon known as the "demographic transition": the more urbanized a given soci-ety, the lower its fertility rate. As a *general* statistical phenomenon, this transition dates to the end of the nineteenth century, but there is some evidence (from cities such as Geneva and Venice) that wealthy classes in the West limited their reproductive output long before that. "Although here the picture is particularly uncertain and complex, it may be that urban dwellers were the first in large numbers to restrict family size with-in marriage, as well as to shape desired family size to economic circum-stances."[34] Many additional factors must be brought to bear to make Colinvaux's model more realistic. The inherent uncertainty of the pre-industrial urban environment, particularly the high infant-mortality rates, made it hard to calculate even a satisfactory family size. People had to produce extra children as insurance against famine and disease, and in the case of farmers, as potential economic contributors. Moreover, there were collective mechanisms of population control:

> Preindustrial western Europe exhibited one striking and aberrant character-istic. While population did tend to grow in the presence of abundant land,

the rate of increase always remained moderate. The fertility rates, lower than in other societies, indicate the presence of preventive checks to births. These checks were communal rather than individual, and amounted to a European system of social control of fertility. The most common mode of control in western Europe was to impose socio-economic conditions on marriage: a tenancy or guild membership for the groom, an appropriate dowry for the bride. As a result, people were often forced to marry late and many remained single throughout life because they could not achieve an independent situation.[35]

The changing role of women in medieval society is another factor that must be added to Colinvaux's model. Recent studies of the demographic transition in modern times make it increasingly clear that a widening of women's niches is as important as urbanization in inducing this bifurcation in the Third World. Women's access to education, contraceptives, and jobs (that is, any expansion beyond the narrow niche of "breeder"), as well as increased decision-making power in the process of family planning, is a prerequisite for the transition.[36] To the extent that women are forced to exist within narrow niches, gender distinctions are very much like class or caste distinctions. That is, reproductive strata are also hierarchical structures, only on a smaller scale, since familial hierarchies exist within socioeconomic ones.

In the previous chapter we noted that hierarchy building consists of two distinct operations, a homogenization performed by a sorting process, followed by a consolidation through coding into legal, religious, or other formal regulations. This is not, of course, a strictly sequential process: in practice, even after a code has been established, new sorting operations continue, alongside or even against the regulated routines. Reproductive niches (or strata) may be seen as the result of such a hierarchy-building process. The initial homogenization is performed on materials supplied by the biological substratum. Some of the traits that are genetically determined in a simple way (raw-milk digestibility, malarial resistance) exist as sharp dichotomies (an individual either possesses the trait or does not), while traits determined by the interaction of multiple genes (or pairs of alleles) form a more or less continuous statistical distribution. The ability to bear children is of the first type, while most of the secondary sexual characteristics (the ones used to define gender roles) are of the second type. Consequently, with respect to the important category of secondary sexual characteristics, genetic materials create two fuzzy statistical distributions (one for males, the other for females) with an *area of overlap*.[37]

When we compare these overlapping fuzzy sets with cultural definitions of gender, in which reified essences such as "rationality" or "emotionality" are sharply dichotomized, we can be sure that a homogenizing operation has taken place. For example, women have traditionally been denied fighting (or even self-defense) skills. In comparison to biologically vital functions such as giving birth and taking care of children (as well as making biomass edible, by grinding, soaking, cooking, and fermenting), fighting may not seem so important, at least not before state-directed wars of conquest began to yield rich spoils. But fighting skills were crucial; their exercise gave people access to certain roles (the warrior) that were sources of prestige and status. Simply in terms of physical strength, women at the upper end of the scale, falling in the area of overlap, would have been superior fighters in comparison to men located at the bottom end of the male scale of physical strength, and yet these "genetically endowed" potential female warriors were excluded from the prestigious role.[38] Moreover, because physical strength can be amplified by training, exclusion meant that the overlap area was *artificially* reduced in size:

> Biology can feed back onto biology through social distinctions: for hormonal reasons, women, on the average (but only on the average), have a different proportion of muscle to fat than men, and this has the consequence that women, on the average (but only on the average), can exert somewhat less physical force on objects. The division of labor between men and women and the division of early training, activity and attitude cause a very considerable exaggeration of this small difference, so that women become physically weaker than men during their development to an extent far in excess of what can be ascribed to hormones.[39]

In medieval Europe, as the historian Edith Ennen has shown, this exclusion from the role of warrior preserved the age-old function of "guardianship" as the exclusive domain of the father or other male member of a patriarchal family. In a sense, the function of this institution (and other related ones) was to control the flow of genes, by means of asymmetrical regulations regarding infertility, infidelity, and ownership of offspring. It is important, however, not to view reproductive strata as static entities, but to focus instead on the dynamics of their defining borders. Ennen writes of the shifting borders of medieval women's roles:

> In the history of women in the Middle Ages there are constants and changes — and there is permanence within the changes. The most powerful

constant: woman as the rich heiress, woman as bearer of successors and heirs. This is true for monarchs and peasants, nobles and burghers. The higher the rank, the more important this "function", the value of which, for the fertile and the pregnant woman, is calculated in money terms in the *werengeld*-regulations of the Frankish *leges* [Germanic tribal law]. The survival of the dynasty depends on her.[40]

Ennen goes on to point out other constants, most importantly, the preservation of the function of guardianship. But Ennen also observes that women's niches were considerably broadened by the advent of urban life and by the slow replacement (in northern Europe) of Germanic law by Christian codes. Prior to this millennium, a marriage contract was entered into by the groom and the woman's guardian; by the year 1030, a woman's consent was required in England. By the twelfth century, the legal principle of marriage by consent was fully established, and imposed marriages were barred, at least in theory.[41] In many cases, of course, family politics still determined whom daughters would marry, since advantageous marriages were one of only a few means for a family to rise socially, but some medieval women did acquire a degree of freedom in choosing a husband.

In medieval towns women's niches were widened in a variety of ways. Women acquired a relatively high degree of commercial independence (in fact, women were more thoroughly excluded from commerce in the nineteenth century than they were in the late Middle Ages[42]), and benefited from changes in the law of matrimonial property as well as in inheritance laws with respect to wives and daughters. Males and females also became equal in citizenship rights, although not in political participation:

In this way [through improved legal status and hereditary rights] women gained a share of civic freedom. In many civic legal codes, e.g. that of Bremen dating from 1186 and of Stade from 1209, the husband and wife are both explicitly mentioned in the important article which states that any person who lives in the town under municipal law for a reasonable period is free. Women swear the civic oath and are entered in the register of citizens. The wife's share of the civic rights of her husband continues in full after his death.... However, the sources do not indicate that women played any part in the gaining of these freedoms, and those who fought for them were not concerned with the emancipation of women in the modern sense. The medieval concept is not based on the notion of a personal sphere of freedom; it is seen in corporate terms, and it is the freedom of the citizenry as a whole, the town community, that is pursued.[43]

Medieval European towns were not only isolated as ecosystems (heat islands and food-web islands) but their walls made them islands in a cultural sense, places where certain privileges could be exercised, where the old feudal restrictions could be relaxed, where new niches (e.g., a middle class) could be created. Unlike individual serfs who were bound to a given manor and its landlord, urban citizens had no such individual obligations, although cities as a whole did owe dues to bishops, counts, or kings. The relative autonomy of towns, which varied from place to place, tended to be reflected in the institutional norms and rules that slowly accumulated within their ramparts. If after some period of residence a town adopted a runaway serf, these institutional norms replaced his or her own allegiance to a lord, and this made the medieval town "a veritable machine for breaking up old bonds."[44] This does not mean, of course, that rural immigrants were not drawn almost immediately into other pyramidal structures. In Braudel's words, "the peasant who uprooted himself from his land and arrived in the town was immediately another man. He was free — or rather he had abandoned a known and hated servitude for another, not always guessing the extent of it before hand."[45]

The urban intensification that peaked by the late thirteenth century created many opportunities for such escapes. While in 1050 a runaway peasant had nowhere to go, since towns were several days from each other, by 1300 most towns were only one day apart. More importantly, while in 1050 towns were surrounded by forbidding forests which acted as barriers to migration, by 1300 these forests were beginning to disappear.[46] But what was beneficial from the perspective of migrating peasants was potentially catastrophic for the urban centers themselves. In two and a half centuries, towns and their supply regions had grown at the expense of the biological meshwork within which they had evolved. The ecosystem was greatly homogenized: many parts of the forest had been cleared and either converted into agricultural land or simply destroyed and used for fuel or construction materials. As one author puts it, urban expansion was bought on credit, using as collateral the continent's natural resources. After 1300, nature foreclosed and Europe faced its first ecological crisis of the millennium. Prior to the fourteenth century, most famines were localized, which meant that regions whose agricultural production failed could import biomass from nearby areas. But after 1300, general famines became common, one of the most severe of which struck in 1315 and lasted several years.[47]

Deforestation of mountain slopes led to erosion and the loss of fertile soil. Although some of this soil accumulated in the valleys below, increasing their fertility, deforestation intensified the frequency of floods, leading to further soil loss and destruction of crops. This happened, for instance, in certain regions of the Upper Rhine Valley.[48] Soil loss due to careless exploitation of the forests' resources, particularly the transformation of steep slopes into agricultural land, has been a constant threat to urban centers throughout history. In fact, some historians postulate that urban life began in Egypt and Mesopotamia precisely because the land there was flat and hence not subject to erosion and soil loss. They calculate that most other urban civilizations were able to pass their genes for only seventy generations before they ran out of soil.[49] Even though methods of preventing erosion were known from the times of the ancient Phoenicians (terracing techniques, for example), many urban hierarchies in the past failed to implement such knowledge. This is another example of the practical limits of bounded rationality, and proof that, although some material and energy flows can be "socialized" (i.e., submitted to cultural control), *in practice* many are not.[50]

In addition to deforestation, the fourteenth-century ecological crisis involved disruptions to the simplified (hence unresilient) ecosystems with which cities and their regions had replaced the forest. By shortening food chains, human populations acquired control over nutrient cycles. For instance, cattle and certain crops went hand in hand: the manure of the cattle, which were raised on cereals, could be plugged back into the system as fertilizer, closing the nutrient cycle. In itself, this tightening of the cycles was good. Indeed, ecosystems spontaneously shorten their nutrient cycles as they complexify. A highly complex system such as a rain forest runs its nutrients so tightly, via elaborate microflora and microfauna in the tree roots, that the soil is largely deprived of nutrients. This is one reason why the destruction of rain forests is so wasteful: the soil left behind is largely sterile. The temperate forests of Europe, on the other hand, do run their nutrient cycles through the soil, and there deforestation leaves a valuable reservoir behind. But when Europeans replaced this ecosystem by taking control of the cycles themselves, unforeseen glitches disrupted the system. For example, as some agricultural lands specialized, and cattle were sent to the highlands to graze, the manure cycle was broken, leading to a loss of soil fertility.[51]

Components of the ecosystem which lie outside social control, such as the climate, also contributed to the ecological crisis. Worldwide cooling trends seem to have afflicted the fourteenth and seventeenth centuries. Braudel notes that even civilizations at great distances from one another (e.g., Europe and China) may have been connected by global climate changes that affected the yield of their harvests and hence the fates of their populations. There is some evidence that the cycles of population growth and decline in the Far East and the Far West were synchronized before the eighteenth century; given the relatively low intensity of commercial contact between East and West, global climate rhythms would seem to be the missing link:

> A general cooling-down process occurred in the Northern hemisphere in the fourteenth century. The number of glaciers and ice-floes increased and winters became more severe. One historian suggests that the Vikings' route to America was cut off by dangerous ice at the time. Another thinks that some dreadful climatic drama finally interrupted European colonization in Greenland, the evidence being the bodies of the last survivors found in the frozen earth.... Similarly the "little ice age" ... during Louis XIV's reign was more a tyrant than the Sun King. Everything moved to its rhythm: cereal-growing Europe and the rice-fields and steppes of Asia.... All this gives additional meaning to the fluctuations of material life, and possibly explains their simultaneity. The possibility of a certain physical and biological history common to all humanity before the great discoveries, the industrial revolution or the interpenetration of economies.[52]

There was another component of urban ecosystems that defied hierarchical control by human cultures and linked the fates of East and West: contagious disease. As we saw, urban ecosystems on both sides of Eurasia (and in many places in between) were epidemiological laboratories where animal diseases evolved into human ones, and where the density of population was intense enough to make the disease endemic, that is, to allow it to subsist in more or less stable coexistence with its human hosts. Many of the childhood diseases that afflicted medieval Europe had been "manufactured" one or two millennia earlier in the four separate "laboratories" that had emerged by classical times (the Mediterranean, the Middle East, India, and China). Smallpox, for instance, may have been brought to the Roman Empire by soldiers returning from a campaign in Mesopotamia.[53] Although each of these centers evolved separately for a while, as the intensity of trade (or warfare) between them intensified, they became interconnected.[54]

The long caravans that continuously traversed the Silk Road and the intense maritime commerce across the Indian Ocean had emerged as the main communication channels connecting the different disease pools. Microorganisms traveled with silk and other goods through these channels, which were sustained by military power, habit, and routine. The accelerated urbanization of Europe a thousand years later and the consequent establishment of regular land and sea routes for commerce had a similar effect at a smaller scale, joining the cities along the Mediterranean coast with the brand-new cities in the north into a single disease pool.[55] These homogenizations of the microscopic component of urban ecosystems had a beneficial effect: had the disease pools remained isolated, any contact between them would have unleashed explosive epidemics.

However, urban populations were not alone in fostering endemic diseases. Wild animal populations, too, harbored colonies of microbes, and contact between these animals and humans could have catastrophic results. That is what happened in 1346, when the bubonic plague was unleashed on Europe. The plague bacillus (*Pasteurella pestis*) had become endemic among underground populations of rats and fleas at the foothills of the Himalayas. The expansion of the Mongol Empire, which converted the old low-intensity trade routes into a complex network of caravansaries extending into the northern Eurasian steppes and connecting China with Europe, had created new disease channels, both for humans and for rats:

> What probably happened between 1331 and 1346 . . . was that as plague spread from caravanserai to caravanserai across Asia and Eastern Europe, and moved thence into adjacent human cities wherever they existed, a parallel movement into underground rodent "cities" of the grasslands also occurred. In human-rat-flea communities above ground, *Pasteurella pestis* remained an unwelcome and lethal visitor, unable to establish permanent lodgment because of the immunity reactions and heavy die-off it provoked among its hosts. In the rodent burrows of the steppe, however, the bacillus found a permanent home. . . . Before the Black Death could strike as it did [in Europe], two more conditions had to be fulfilled. First of all, populations of black rats of the kind whose fleas were liable to carry bubonic plague to humans had to spread throughout the European continent. Secondly, a network of shipping had to connect the Mediterranean with northern Europe, so as to be able to carry infected rats and fleas to all the ports of the Continent. Very likely the spread of black rats into northern Europe was itself a result of the intensification of shipping contacts between the Mediterranean and northern ports.[56]

Hence, the same intimate contacts that had made medieval cities into a single disease pool, which prevented their contagious diseases from becoming epidemic, now worked against them by allowing cross-border contact between urban populations and disease-carrying rats and fleas, which spread the plague rapidly across Europe. According to William McNeill, it took about 100 to 133 years (five or six human generations) for the plague to become endemic.[57] Nevertheless, because endemic equilibrium may be cyclical, localized epidemic outbreaks of varying intensity continued until at least the eighteenth century. In the first massive outbreak (1346–1350), about a third of the European population was consumed by the plague. Subsequent waves were almost as lethal, and it seemed as if urban and rural Europe were being digested from within by weeds (rats, fleas) and their microparasites.

The social consequences of the intensified mortality rates were numerous. The peasantry and working classes benefited in the sense that the survivors found themselves in a world with acute labor shortages, not to mention the fact that the survivors inherited the possessions of those eaten by the plague. Wages increased, broadening workers' niches significantly. These might be described as Pyrrhic benefits, however, since the urban and rural poor sustained the vast majority of casualties. The rich would abandon a city at the first signs of epidemic, while "the poor remained alone, penned up in the contaminated town where the State fed them, isolated them, blockaded them and kept them under observation."[58] Not only the inhabitants but the cities themselves "died," since many of those who played key roles in government and commerce fled and key urban functions (business and legal activities, religious services) ceased operating.

Despite a general dereliction of duty, government hierarchies did respond to the challenge, through a variety of methods, including quarantines, surveillance, inhalants, disinfection, blocked roads, close confinement, and health certificates.[59] Planned response, however, remained ineffectual, not only because of the limitations of bounded rationality, but also because the cause of the plague (a bacillus) and its method of contagion (rats, fleas, humans) defied human comprehension until the late nineteenth century. Nevertheless, in the eyes of the survivors, secular authorities had at least made an effort to fight back, while ecclesiastical hierarchies had remained powerless to cope with the emergency. In the aftermath, the authority of the church emerged damaged (anticlericalism intensified) while secular hierarchies were strengthened.[60] In the end, however, it was not any planned response that stopped the plague, but a trial-and-error accommodation to it.[61]

There were other social consequences of the plague. After each successive epidemic wave had passed, the gene flow between classes increased in intensity. Cities found themselves depopulated and lowered their standards for citizenship. Venice, normally very closed to foreigners, now granted free citizenship to anyone who settled there for a year.[62] Social mobility increased, as surviving elites needed to replenish their ranks with fresh blood. Relationships among cities altered because of the enormous demographic shifts wrought by the plague. The eventual emergence of Venice as the core of the Network system was in no small measure a consequence of those demographic changes.[63]

The Black Death struck a European population that was already afflicted by an ecological crisis of its own making. Although the deforestation that precipitated this crisis was the product of intensified urbanization, we should distinguish a variety of roles played by different types of cities. The cities of the Central Place system—that is, landlocked hierarchies of towns of different sizes—cleared their forests for farmland, for the reservoir of nutrients that the temperate forests' soil contained. The gateway ports of the Network system, on the other hand, marketed the indigestible biomass of the forest (wood) as fuel or construction material for ships. More accurately, the various regions that gave birth to the maritime metropolises of Europe rose to prominence by exploiting three different reservoirs: timber, salt, and fish.[64] While some Central Place hierarchies exterminated their forests with almost religious zeal (in some cases using specialized monks who thought of every acre cleared of demon-infested forest as an acre gained for God[65]), Network-system gateways had a more managerial attitude toward their reservoirs.

There were, of course, mixtures. Some Central Place cities, such as Paris, housed hierarchies that viewed their forests as renewable resources. French forests were stabilized in the sixteenth and seventeenth centuries, partly by decree (the great ordinance of 1573 and the measures taken by Colbert) and partly because the remaining forest soils were too poor to exploit.[66] Nevertheless, there were important differences between metropolises and capitals as ecosystems which influenced their relationships to the flow of biomass, edible and inedible. Many of the seaports—and certainly all the ones that served as core of the Network system before the nineteenth century (Venice, Genoa, Amsterdam)—were *ecologically deprived* places, incapable of feeding themselves. In this sense, they were all like Amalfi, a small Mediterranean port whose hinterlands were largely infertile, but that at the turn of the millennium had served as a gateway to the dynamic markets of Islam and had played a key role in the reawakening of Europe.

> Like Amalfi in its hollow among the mountains, Venice, scattered over sixty or so islands and islets, was a strange world, a.refuge perhaps but hardly a convenient one: there was no fresh water, no food supply—only salt in abundance.... Is this an example of the town reduced to bare essentials, stripped of everything not strictly urban, and condemned, in order to sur-

vive, to obtain everything from trade: wheat or millet, rye, meat on the hoof, cheese, vegetables, wine, oil, timber, stone—and even drinking water? Venice's entire population lived outside the "primary sector"...[her] activities all fell into the sectors which economists would nowadays describe as secondary and tertiary: industry, commerce, services.[67]

The same is true of Genoa, which was the financial capital of sixteenth-century Europe: the city arose on a small strip of land surrounded by mountains barren of trees and even grass.[68] The extreme poverty of the lands on which the Italian maritime metropolises were built was partly due to the soil depletion caused by previous intensifications. In many regions in and around the Mediterranean where production had been intensified a thousand years earlier to feed the cities of the Roman Empire, erosion had long since removed the fleshy soil and exposed the underlying limestone skeleton. According to some historians, only the soil north of the Po Valley had been spared this destruction, and these were the lands that later fed medieval Europe. The regions that had been the stage of barbaric invasions and war after the fall of the Roman Empire had also recovered their fertility by medieval times, since military turbulence made continuous intensified agriculture impossible.[69] But the land on which towns like Venice, Genoa, or Amalfi grew still bore the scars of careless intensification. Thus, although many cities in the fourteenth century (e.g., Florence) were already importing grain from far away, towns such as Venice and Genoa were, from the start, *condemned to trade* to maintain their lifeline.

There are other interesting differences between Central Place and Network cities in this respect. Although the former were better endowed ecologically, even for them continuous growth entailed intensification and hence depletion. At some point, either trade or invasion became necessary to tap into the nutrient reservoirs of ever more distant soils. While cities belonging to territorial states invaded other peoples' lands, gateway ports penetrated their markets. In other words (and allowing for complex mixtures), landlocked capitals took over fertile lands, at times giving birth to a landlocked colonial city on foreign soil and redirecting the flow of biomass to the motherland. Metropolises, on the other hand, took over strategically located albeit barren pieces of rock in the middle of the ocean, to control the trade routes that connected Europe to lucrative foreign markets. As Braudel says, "In order to control the large expanses in question, it was sufficient to hold a few strategic points (Candia, captured by Venice in 1204; Corfu, 1383; Cyprus, 1489—or indeed Gibraltar, which the British took by surprise in 1704, and Malta, which they captured in

1800) and to establish a few convenient monopolies, which then had to be maintained in good working order—as we do machines today."[70]

From these strategic places a naval power could control the Mediterranean (and the markets of the Levant) and, hence, the trade lifeline of the region. From likewise ecologically poor strongholds on foreign coasts, or from foreign gateway cities, European metropolises acquired control of faraway markets in India, China, and the Levant. From these entry points, they captured and redirected a continuous flow of luxury goods (spices, for example), with perhaps negligible nutritional value but capable of generating extraordinary profits. It is true that some gateways also engaged in the colonization of nearby lands for their soils, as when Venice took control of the Italian mainland around it (including the towns of Padua, Verona, Brescia, and Beragamo) in the early 1400s. But even there, the land was soon used not to feed the city, but to raise cash crops and livestock for the market. Amsterdam, another ecologically poor gateway port, and its sister cities in the United Provinces shaped their limited holdings of fertile land into an efficient agricultural machine, though it, too, was oriented toward external markets.[71] In many respects, these Network cities were not tied to the land and exhibited the kind of weightlessness, or lack of inertia, that we associate with transnational corporations today. Is it any wonder that maritime metropolises such as Genoa or Venice (as well as those regional capitals closely connected to them, such as Florence or Milan) were the birthplace of many antimarket institutions?

Braudel invites us to view the history of the millennium as three separate flows moving at different speeds. On one hand, we have the life of the peasant population, more or less chained to the land, whose customs change with the viscosity of lava. Corn, which fed Europe, and rice, which fed China, were tyrants that forced on the peasantry a rigid adherence to well-worn habits and routines and a closed cycle of production. This is what Braudel calls "material life," the know-how and traditional tools, the inherited recipes and customs, with which human beings interact with plants to generate the flow of biomass that sustains villages and towns. This body of knowledge resists innovations and hence changes very slowly, as if history barely flowed through it. One historian suggests that one needs observational timescales a millennium long to understand the agrarian structures of Italy.[72] The peasant masses are, in a sense, like the assemblage of flora at the base of natural ecosystems, an immobile engine that creates the energy which makes everything around them move.

Next comes the world of markets and commercial life, where the flow of history becomes less viscous. Braudel calls market towns "accelerators of all historical time."[73] Although peasants sometimes came to the city market of their own accord, more often than not they were forced to come, and to that extent we may say that towns fed on them, or parts of them, much as an herbivore does. So above the bottom layer of material life

> comes the favoured terrain of the market economy with its many hori-
> zontal communications between the different markets: here a degree of
> automatic coordination usually links supply, demand and prices. Then
> alongside, or rather above this layer, comes the zone of the anti-market,
> where the great predators roam and the law of the jungle operates. This —
> today as in the past, before and after the industrial revolution — is the real
> home of capitalism.[74]

This is the layer of *maximum mobility*, where large amounts of financial capital, for example, flowed continuously from one highly profitable area to another, defying frontiers and accelerating many historical processes. In summary, according to Braudel, the European economy comprised three spheres or layers: the inertial peasant layer, which was the source of biomass flow; the market economy, which set surpluses into motion by means of the flow of money; and the antimarket, where money detached itself from biomass, becoming a mobile mutant flow capable of investing in any activity that intensified the production of profits. This ultimate layer may be properly called "predatory" to emphasize its noncompetitive and monopolistic (or oligopolistic) nature. Antimarkets, of course, coexisted with other predators (or as McNeill calls them, "macroparasites"[75]), such as central states and feudal hierarchies, which also derived their sustenance by tapping into the energetic flows produced by others, via taxes, rents, or forced labor.

These hierarchies (all urban in the case of medieval Italy) sometimes metamorphosed from one type of macroparasite into another. Wealthy merchants and financiers, for instance, would retire from business and buy land, sacrificing their mobility in hopes of acquiring access to the aristocracy and the opportunity to spread their genes across class barriers. Noble landlords, on the other hand, would sometimes take advantage of their monopolies of soil, timber, and mineral deposits to play antimarket roles, albeit lacking the rationalization and routinization that characterized big business. More often than not, however, these noblemen collaborated in the transfer of surpluses from agricultural regions.

As Europe's urban ecosystems expanded and multiplied their interconnections with one another, they became not only a single disease pool but a single economy as well. Soon the simple relationship between a city and its surrounding supply zone of small villages was left behind (at least outside the lower ranks of Central Place hierarchies), and many large towns began to draw their nutrients largely from a single, vast source, replicating on a huge scale the original parasitic relationship the individual cities had with their countrysides. In other words, during the sixteenth century Europe began colonizing itself, transforming its eastern regions (Poland and other territories east of the Hamburg-Vienna-Venice axis) into its supply zone. As with all such peripheral regions, their relationship to the core that exploited them was mostly negative: their own market towns lost vitality, hostility to innovation increased, and barriers between classes hardened. The result was that, unlike small towns in the middle zone which could trade with one another and eventually shake their subordinate position, these peripheral areas were condemned to a permanent state of backwardness.

In the case of Eastern Europe, its reduction to colonial status was brought about by the actions of several hierarchies: the local landlords, who intensified their macroparasitism to an extreme (six days a week of forced labor was not uncommon for peasants), and wholesalers in cities such as Amsterdam who preyed on the landlords themselves, manipulating supply and demand through warehousing and advanced purchases from producers.[76] As this internal colonization was taking place, Europe was beginning to develop a core-periphery relationship on an even larger scale, this time at a global level. Spain and Portugal, whose soils had not recovered from the intensification of the Roman Empire, spearheaded the conquest of lands across the Atlantic, the conversion of America into a continent-wide supply zone.

Medieval cities had attempted a first round of foreign colonization centuries earlier, at the time of the Crusades, but this earlier effort had lacked staying power. Despite the hundreds of thousands of Europeans who had been mobilized for the invasion of the Holy Lands, Europe's colonies abroad (Edessa, Antioch, Tripoli, Jerusalem) had promptly returned to Islamic control. Much as population density was the only means to maintain the domination of urban over forest ecosystems (drops in population allowed the return of banished plants and wolves), here, too, density was needed to sustain a European presence on foreign soil. And yet, as one historian puts it, despite the original massive transfer of people, Europe "lost the propagation game."[77] In addition, there was another great biological barrier to the success of the Crusades—microorganisms:

When the Crusaders arrived in the Levant, they had to undergo what British settlers in the North American colonies centuries later called "seasoning"; they had to ingest and build resistance to the local bacterial flora. They had to survive the infections, work out a modus vivendi with the Eastern micro-life and parasites. Then they could fight the Saracens. This period of seasoning stole time, strength and efficiency, and ended in death of tens of thousands. It is likely that the disease that affected the Crusaders the most was malaria.... Crusaders from the Mediterranean ... had brought with them a degree of resistance to malaria.... Unfortunately for [them], a person immune to one kind of malaria is not immune to all, and immunity to malaria is not long-lasting.[78]

Genes that provide resistance to malaria (the sickle-cell and beta-thalassemia genes) existed in the southern European gene pool, but they were rare in the north. Consequently, Crusaders from France, Germany, and England were devoured from within by the particularly virulent malarial strains endemic in the Middle East. When Europe began colonizing faraway lands four hundred years later, she confronted an entirely different situation. Now her childhood diseases, particularly smallpox and measles, fought on her side. As McNeill says, these were a "biological weapon urban conditions of life [had] implanted in the bloodstreams of civilized peoples."[79] In fact, whenever encounters took place between human populations that had not been in close contact with one another and only the invaders possessed "civilized" diseases, the affair resembled a gigantic food chain in which one mass of humans ingested the other:

First, the structural organization of neighboring communities was broken down by a combination of war (cf. mastication) and disease (cf. the chemical and physical action of stomach and intestines). Sometimes, no doubt, a local population suffered total extinction, but this was not typical. More often, the shattering initial encounters with civilization left substantial numbers of culturally disoriented individuals on the land. Such human material could then be incorporated into the tissues of the enlarged civilization itself, either as individuals or as small family and village groupings.[80]

As Europe began reaching out into the world to create new supply regions, European diseases visited near-extinction or, alternatively, decimation on the indigenous populations. In one of the first successful attempts at colonization (the Canary Islands), the local peoples (the Guanches) were driven to the brink of extinction, mostly by the invaders' diseases. Today a few Guanche genes remain in the Canaries' gene pool, along with a few words and nine sentences from their original language.[81] The rest was annihilated. On the other hand, in what proved to be the most successful and long lasting colonial enterprise, the conversion of the American continent into a huge peripheral zone to feed the European core, only some areas (the United States, Canada, Argentina) witnessed the wholesale replacement of one gene pool by another. In the rest of the Americas, entire communities were instead culturally absorbed. Like those insects that first regurgitate a soup of enzymes to predigest their food, the conquerors from Spain killed or weakened their victims with smallpox and measles before proceeding to Christianize them and incorporate them into the colonial culture.

Earlier attempts at colonizing the New World had failed partly because of a lack of "predigestive enzymes." The Norse, who tried to colonize this continent earlier in the millennium failed because their motherland (Greenland) was "so remote from Europe that they rarely received the latest installments of the diseases germinating in European centers of dense settlement, and their tiny populations were too small for the maintenance of crowd diseases."[82] The new wave of invaders from Spain not only were in direct contact with the epidemiological laboratories that "manufactured" these biological weapons, they were the fleshy component of the disease factory. The local Amerindians, on the other hand, though densely populated enough to sustain endemic relations with parasites, lacked other components of the laboratory: the livestock that coexisted with humans and exchanged diseases with them.[83]

Overall, the effects of the encounter between epidemiologically scarred Europe and virgin America were devastating. The total population of the New World before the Conquest was by some estimates as high as 1 hundred million people, one-third of whom belonged to the Mexican and another third to the Andean civilizations. Fifty years later, after its initial encounter with Cortés, the Mexican population had decreased to a mere 3 million (about one-tenth of the original).[84] After the initial clash in Mexico in 1518, smallpox traveled south, reaching the Inca empire by 1526, long before Pizarro's troops began their depredations. The disease had equally drastic consequences, making it much easier for the conquerors to plunder the Incas' treasures and resources. The measles followed smallpox, spreading through Mexico and Peru in the years 1530–1531. Other endemic diseases such as diphtheria and the mumps soon crossed the ocean, and even some of the epidemics that still afflicted Europe (e.g., typhus and influenza) may have also leaped this ancient seawater barrier: the globe was beginning to form a single disease pool.[85]

The cultural advantages that the Spanish enjoyed (horses, very primitive firearms, metal armor) would have been quite insufficient for the task of conquering a densely inhabited territory. Large animals and loud weapons had, no doubt, a powerful psychological effect on the native population. But after the first encounters, during which the indigenous warriors saw their stone weapons pierce through European armor and horseflesh and witnessed the inefficiency of the Spaniards' inaccurate, single-shot muskets, these cultural advantages would have dissipated. But because the majority of the native inhabitants died from disease, draining the reservoirs of skills and know-how that sustained their culture, that meager advantage sufficed. Culture certainly played a role here, but it was not the most important. Cultural materials flowed together with genes and biomass (not all of it human) across the Atlantic, and it was the whole complex mixture that triumphed.

An entire continent was in this way transformed into a supply region for all three spheres of the European economy: material life, markets, and antimarkets. Sugar and other inexpensive foodstuffs for the masses would soon begin flowing in large quantities from the colonies and plantations to the homeland. A variety of raw materials to be sold in her markets also flowed home. Finally, an intense flow of silver (and other precious metals) provided fuel for European antimarkets and for the European monetary system as a whole.

We saw above that while some cities took over alien lands other cities tapped into foreign resources by manipulating markets. Unlike the process of colonizing a territory, a mostly biological affair, penetrating foreign markets (such as the huge Indian or Chinese markets, which rivaled those of Europe until the eighteenth century) involved large quantities of metallic money. Silver (rather than infectious diseases) played the role of "predigestive enzyme" here. Thanks in part to the steady flow of metal from American deposits, the European monetary system "was projected over the whole world, a vast net thrown over the wealth of other continents. It was no minor detail that for Europe's gain the treasures of America were exported as far as the Far East, to be converted into local money or ingots in the sixteenth century. Europe was beginning to devour, to digest the world."[86]

Central Place capitals such as Madrid, Network-system metropolises such as Amsterdam, and hybrids such as London used their own biological or mineral materials to dissolve foreign defenses, break apart loyalties, weaken the grip of indigenous traditions. After gaining entry onto foreign soil this way, a massive transfer of people, plants, and animals was necessary to establish a permanent European presence. In some areas of the world, particularly those that had been used as gateways to exploit foreign markets, the new colonies would fail much as those established during the Crusades had. But in other parts, Western colonizers would indeed win the propagation game and, with it, access to the most fertile and productive lands of the planet.

Species and Ecosystems

We would do well to pause now for a moment to consider some of the philosophical questions raised by the flow of genes and biomass, as well as by the structures that emerge from those flows. As I argued in the previous chapter, there is a sense in which species and ecosystems are the product of structure-generating processes that are basically the same as those which produce the different types of rocks that populate the world of geology. A given species (or, more accurately, the gene pool of a species) can be seen as the historical out-come of a sorting process (an accumulation of genetic materials under the influence of selection pressures) followed by a process of consolidation (reproductive isolation), which gives a loose accumulation of genes a more or less durable form by acting as a "ratchet device." The most familiar form of repro-ductive isolation considered by biologists has an external cause: geographical changes in the habitat where reproductive commu-nities belonging to that species live. For instance, a river may change its course (over many years) and run through the middle of a previously undivided territory, making contact between members of a reproductive community difficult or impossible. In that situation, the two halves of the community will start to accumulate changes indepen-dently of each other and hence begin to diverge, until the day when mating between their respective members becomes (mech-anically) impossible, or produces only sterile offspring.

136

However, the process of reproductive isolation (and thus, of *speciation*) may be more complex than that; in particular, it may have *internal* causes as well as external ones. One well-studied example of an internal cause is the "specific mate recognition system," or SMRS.[87] This is the system of traits and signals (which can be behavioral or anatomical, or both) that members of a sexually reproducing species use to recognize potential mates. Genetic changes that affect the SMRS (mating calls, courtship rituals, identifying marks and decorations, smells) may indeed act as a barrier to interbreeding even if the two divergent daughter species could potentially mix their genes. In this case, sexual selection (that is, selection pressures exercised on an individual by its potential mates) can cause a small initial difference to be amplified into a major barrier to the exchange of genes and, hence, result in the creation of a new species.[87]

Thus the flow of genes (which one might imagine as potentially continuous) becomes encapsulated via these isolating barriers into separate packets, each defining a different stratified system. However, there is a risk of

exaggerating the strength of these barriers, particularly if we pay attention only to the world of relatively large animals, to which we belong.
Indeed, other living creatures may not be as genetically "compartmentalized" as we are. Many plants, for example, are able to hybridize with
plants of other species (that is, the isolating barriers retain a measure of
permeability), while many microorganisms freely exchange genes with
other species during their lifetimes. (As we shall see, this seems to be the
way many of the bacteria that cause infectious diseases have acquired
resistance to antibiotics.) In short, the flow of genes in the biosphere as
a whole may not be as discontinuous (as stratified) as one would imagine
by looking at large animals alone. In fact, in some special circumstances,
even animals in total reproductive isolation may exchange genetic materials via inheritable viruses (called *retroviruses*).[89]

Taking all this into account, the picture of evolutionary processes that
emerges resembles more a meshwork than a strict hierarchy, a bush or
rhizome more than a neatly branching tree:

> There is substantial evidence that organisms are not limited for their evolu-
> tion to genes that belong to the gene pool of their species. Rather it seems
> more plausible that in the time-scale of evolution the whole of the gene pool
> of the biosphere is available to all organisms and that the more dramatic
> steps and apparent discontinuities in evolution are in fact attributable to very
> rare events involving the adoption of part or all of a foreign genome. Organ-
> isms and genomes may thus be regarded as compartments of the biosphere
> through which genes in general circulate at various rates and in which indi-
> vidual genes and operons may be incorporated if of sufficient advantage.[90]

Even with this added complication, the two abstract machines discussed in the previous chapter (one generating hierarchies, the other
meshworks) are adequate to account for living structures, particularly if
we make allowance for varying mixtures of the two types. However, I
would like to argue that there is another abstract machine involved in the
production of biological entities which has no counterpart in the geological world, therefore distinguishing species from sedimentary rocks. This
other abstract machine, however, may be found in other nonbiological
realms (in human culture, for instance) and therefore does not constitute
the "essence" of living creatures.

Darwin's basic insight was that animal and plant species are the cumulative result of a process of descent with modification. Later on, however,
scientists came to realize that *any variable replicator* (not just genetic replicators) coupled to *any sorting device* (not just ecological selection pres-

sures) would generate a capacity for evolution. For instance, in the 1970s, the computer scientist John Holland devised a small computer program that self-replicated by following a set of coded instructions and transmitting a copy of those instructions to its progeny. Holland's program did very little other than generate variable replicating copies of itself. However, if a *population* of these replicating programs was submitted to some selection pressure (for example, if the user of the program were to weed out those variants that did not seem an improvement, letting only the more promising variants survive), the individual programs developed useful properties after many generations. This is the basis for Holland's "genetic algorithm," which is widely used today in some computer-based disciplines, as an effective problem-solving device.[91] Richard Dawkins independently realized that patterns of animal behavior (such as birdsongs or the use of tools by apes) could indeed replicate themselves if they spread across a population (and across generations) by *imitation*. Birdsongs are the most thoroughly studied example of these replicators ("memes," as Dawkins calls them), and they do indeed evolve new forms and generate different dialects.[92]

In each of these cases, the coupling of variable replicators with a selection pressure results in a kind of "searching device" (or "probe head") that explores a space of possible forms (the space of possible organic shapes, or birdsongs, or solutions to computer problems). This searching device is, of course, blind (or more exactly, shortsighted), following the key principle of neo-Darwinism: *evolution has no foresight*.[93] (It is, nevertheless, highly effective, at least in certain circumstances.) This probe head is the abstract machine we were looking for, the one that differentiates the process of sedimentary-rock formation from the process that yields biological species. And yet, although the new machine is characteristic of life-forms, the same basic diagram applies to memes and genetic algorithms. It would be incorrect to say that evolutionary concepts are used *metaphorically* when applied to computer programs and birdsongs, but literally when talking about genes. It is true that scientists first discovered this diagram in the world of living creatures, and it may even be true that the living world was the first physical realization of the abstract machine on this planet. However, that does not make the abstract machine any more "intimately related" to DNA than to any other replicator. Hence, it does *not* constitute an "essence" of life, in the sense of being *that which makes life what it is*.[94]

The flow of genes through replication is indeed only a part of what life is. The other part is constituted by the flow of biomass. Individual animals are not just members of a species, but members of a particular

reproductive community inhabiting a particular ecosystem and thus participate in the exchange of energy and materials that makes up a food
web. As with any physical system, the intense flow of energy moving
through an ecosystem pushes it far from equilibrium and endows it with
the ability to generate its own dynamic stable states (attractors). The
same dynamic holds true for the individual organisms evolving within
that ecosystem. Consequently, the space that the probe head blindly
explores is not completely unstructured but already populated by various
types of stable states (static, cyclical, chaotic, autopoietic). This prestruc-
turing of the search space by intensifications of the energy flow may
indeed facilitate the job of the abstract machine (blind as it is). For example, since one possible endogenously generated stable state is a periodic
attractor, which would automatically draw gene activity and gene products into a cycle, the searching device may have *stumbled upon* the
means to generate a primitive metabolism very early on. Further evolutionary complexification may have been achieved as the probe head
moved from attractor to attractor, like so many stepping-stones.

When search spaces (or "adaptive landscapes") were first postulated in
biology in the 1930s, they were thought to be prestructured by a single
equilibrium, a kind of mountain with one peak, which selection pressures
forced the probe head to climb. According to this schema, the top of the
mountain represented the point of maximum fitness, and once a population had been driven there, selection pressures would keep it locked into
this optimal equilibrium. However, recent explorations of adaptive landscapes, using sophisticated computer simulations, have revealed that
these search spaces are anything but simple, that they may comprise
many mountains of different heights (local optima), clustered in a variety
of ways, the valleys and peaks related not directly to fitness but to underlying dynamical stable states. Moreover, once the question of coevolution
is introduced (as when an improvement in a prey's armor puts pressure
on its predator's fangs and claws to further sharpen, which in turn stimu-
lates a thickening of the armor), it becomes clear that interacting species
in an ecosystem have the ability *to change each other's adaptive
landscapes*. (This is just another way of saying that in a predator-prey
arms race there is not a fixed definition of what counts as "the fittest.")[95]

Although the notion of unique stable states did some damage to evolutionary biology (by imposing an oversimplified version of evolution which
disregarded energy flow and the far-from-equilibrium conditions the flow
of energy generates), the idea of the "survival of the fittest" had much
more damaging effects when it was applied to human culture. That mis-
application degenerated almost immediately into Social Darwinism and

the eugenics movement and, later on, inspired the racial cleansing policies of Nazi Germany. Coming as it did after centuries of intense colonialism, Social Darwinism naturally fostered the idea that the Caucasian race
was superior to all others. Of course, in addition to the mistaken notion
of a single, optimal equilibrium, these social movements were nurtured
by the belief that genes determine culture, that is, that there is but a single probe head (whereas, as we just saw, even birds embody at least two).

In reaction to this position, a number of anthropologists (including
Franz Boas, Margaret Mead, and Ruth Benedict) developed during the
first decades of the twentieth century a countertheory that not only gave
human culture its deserved autonomy from genetic determination, but
denied that biological evolution had any effect whatsoever on the development of human societies. According to these anthropologists, human
nature was completely malleable and flexible, and human behavior determined by culture alone. In the short run, "cultural relativism" (as it came
to be known) did us the considerable service of fostering a greater tolerance of cultural differences (a welcome antidote to the racist ideas and
policies of the Social Darwinists and eugenicists), but later on it hardened
into dogma, and in some cases it even degenerated into empty clichés
(such as the slogan "everything is socially constructed").[96]

Fortunately, anthropologists seem to be moving away from dogmatic
positions and developing a new *interactionist* approach, wherein both
organic and cultural evolution are considered simultaneously. One version
of this new approach (the one developed by William Durham) seems particularly close to the view we are exploring here: that both organic and
cultural change involve replicators and that new structures arise by selective retention of variants. Moreover, Durham agrees that this does not
involve a metaphorical use of biological concepts. (He calls this Campbell's rule: the analogy to cultural accumulations is not from organic evolution but from a general model of evolutionary change, of which organic
evolution is but one instance.)[97]

Before describing the five different ways in which genetic and cultural
replicators interact according to Durham, we must first address the question of just what genetic effects we are considering here. Although a few
individual genes have been added to the human gene pool in historical
times (such as the gene that causes sickle-cell anemia but protects its
carriers against malaria), genetic evolution is so much slower than cultural evolution that its influence in human affairs is marginal. As Stephen
Jay Gould points out, "While the gene for sickle-cell anemia declines in frequency among black Americans [since they are not subjected to the
malarial selection pressure], we have invented the railroad, the automo-

bile, radio and television, the atom bomb, the computer, the airplane and spaceship."[98] Thus, the genetic effects we are considering are the organic limitations imposed on us by our own bodies which can be called "human universals" as long as we do not attach any transcendental meaning to this term. (Organic constraints, like cultural constraints, are contingent historical products, though they operate over longer timescales.)

One way in which genetic and cultural replicators interact (or act on one another) is as sorting devices. On the one hand, genes, or rather their bodily (or phenotypic) effects, may act as selection pressures on the accumulation of cultural materials. Durham discusses the example of color perception, and its relationship with color words, partly because its anatomical basis is relatively well known (both the pigment-based system of light absorption in the eye and the processing of sensory input by the brain) and partly because much anthropological research on this subject already exists. Crucial evidence on the "universality" of color perception was gathered in the 1960s by the anthropologists Brent Berlin and Paul Kay in the course of an experiment designed to prove the opposite hypothesis: that each language performs the coding of color experience in a different manner. Berlin and Kay showed a large sample of color chips to subjects belonging to twenty different linguistic communities and asked them to locate in the grid of chips both what the subjects would consider to be the focal point of the referent of a given color word as well as its outer boundaries. On the basis of the linguistic relativity hypothesis (that there is no "natural" way to cut up the spectrum), these researchers expected their experiments to elicit widely scattered focal points and discordant outer boundaries, but instead they recorded a very tight clustering of focal points (and concordance of boundaries) regardless of how many color terms existed in a given native vocabulary. More recent research has supported (and refined) Berlin and Kay's results and has further shown that even though different cultures have accumulated a different number of color labels, the *order that this accumulation follows* exhibits some definite regularities, with terms for "black" and "white" always appearing first, followed by terms for primary colors in certain sequences (red-green-yellow-blue, for example). One possible interpretation is that the first labels that accumulate ("black" and "white") designate broad, composite categories ("dark-cool" and "light-warm," respectively), which slowly differentiate as new labels are added to the repertoire, each one entering the set in a specific and highly constrained fashion. On this basis, Durham has concluded that this is an example of genetic constraints on perception guiding the accumulation of cultural replicators (color words).[99]

Cultural materials, in turn, may act in the opposite direction and influence the accumulation of genes. Unlike the accumulation of color terms,
however, the accumulation of genetic materials happens so slowly as to
be virtually unobservable. Hence, hard evidence is much more difficult to
obtain in this case, and we are forced to discuss hypothetical scenarios
on the basis of indirect evidence, such as that provided by myths. The
example Durham discusses in detail is the gene that allows some Indo-
European races to digest raw milk as adults. First of all, variation for this
gene does exist and is highly correlated with certain cultural patterns.
High prevalences of this gene exist only in populations that today consume
comparatively large amounts of fresh milk and possess ancient mytholo-
gies that both record and encourage adult fresh-milk consumption. In
turn, these genetic and cultural materials are associated with environments of low ultraviolet radiation, where vitamin D and metabolic calcium
are chronically deficient, that is, with environments where fresh-milk consumption can have positive health effects. Durham reviews several possible scenarios that may explain these correlations and concludes that the
most plausible one (as well as the one more consistent with the history
coded into myths) is as follows:

> As genes for LA [lactose absorption] were favored at high latitudes, more
> people could drink milk after weaning, thereby spreading the benefits of
> milk production and improving the local cultural evaluation of the memes
> behind the practice. The increased availability of milk, in turn, would have
> continued the genetic selection of LA genotypes, thereby augmenting the
> frequency of adult lactose absorption, the benefits of milking, the cultural
> preference for milk, and so on in perpetuity. . . . The cycle may have started
> as a continuation of routine infant feeding practices. Early on, the milk of
> dairy animals may have been tried as a supplement to mother's milk,
> increasing the volume of lactation, its duration, or both. By virtue of the
> (initially rare) LA genotypes, some recipients would have maintained lactose
> sufficiency beyond its normal lapse, continuing to drink milk and thereby
> avoiding rickets in their early years. . . . In particularly rachitogenic areas,
> the advantage to fresh milk consumption would have extended into adoles-
> cence and adulthood.[100]

In addition to these two ways of interacting directly with each other, culture and genes may enter into other, more indirect relations. In particular,
Durham points out that once certain cultural materials have accumulated,
they may harden into institutional values, which in turn act as selection
pressures for further cultural accumulations. Hence, some cultural repli-

cators may, in a sense, be *self-selecting*, and this gives them a degree of autonomy in their evolution. Under these conditions, cultural adaptations may come to have relations of enhancement, opposition, or neutrality with respect to genetic adaptations.

Incest taboos are an example of enhancement. Zoologists have convincingly demonstrated that inbreeding has deleterious genetic effects and that many animals have evolved an instinctive avoidance of it. Humans may indeed share this built-in constraint, as studies of aversion to sexual intercourse among adults who were reared together in kibbutzim seem to show. However, as Durham points out, taboo prohibitions are not necessarily the same as avoidance of inbreeding. He observes that "there can be nonincestuous inbreeding (as when sexual intercourse between certain categories of kin is not prohibited) and noninbred incest (as when prohibitions apply between parents and their adopted children)."[101] Given the range of variability of the incest prohibitions, which only partially overlap with inbreeding, Durham concludes that the sets of regulations that constitute the taboo in different societies evolved under cultural selection pressures (although it is possible that instinctive avoidance may have played a role in their accumulation early on in human evolution).

The relative autonomy with which self-selection endows the evolution of cultural replicators allows them to follow a direction that is neutral relative to organic adaptations. For the same reason (i.e., cultural replicators' relative evolutionary autonomy), various aspects of culture may turn out to have maladaptive consequences relative to our biology. For example, many civilizations in the past carelessly intensified the exploitation of their soils, failing to implement available techniques (such as terracing) that could have protected this valuable resource from eroding away. Consequently, those societies inadvertently set a limit on the number of times they could pass their genes down through the generations. (An upper limit of seventy generations existed for most cultures, according to one historian's calculations.) In this case, the bounded rationality of many elites and the prospect of short-term gains promoted the accumulation of habits and routines that, in the long run, destroyed the conditions under which the gene pools of those civilizations could reproduce themselves. Durham also finds these maladaptive cultural materials accumulating in contemporary communities of El Salvador and Honduras, their landscapes "littered with telltale signs of maladaptation. Slopes of forty or fifty degrees ... were being cultivated in perpetuity ... with steadily declining yields. Corn was cultivated in rock outcrops, animals grazed in steep gullies, and the erosive force of tropical rains carried off ever more of the leached and worn-out topsoil."[102] In this case, however, the problem is not the local peasant culture. Rather, the manipulation of land tenure policies by the landed elites and the government's support for export agriculture had imposed these maladaptive conditions on the peasants. From this and other cases, Durham concludes that a major cause of opposition between genetic and cultural replicators is the imposition from above of habits and customs (or living conditions leading to certain habits and customs) that are maladaptive.

However, one must not assume that the power to impose a set of values on a population (and hence to influence the direction of that population's cultural evolution) is always strong enough to eliminate the selective effect of individual choice. (Herein lies another weakness of "cultural relativism": not only does it emphasize the exotic at the expense of the unremarkable, which is where human universals are to be found, but it tends to focus on the norms of a society while ignoring the actual behavior of individual agents, who may or may not always adhere to those norms. Perfect obedience cannot be taken for granted.[103]) According to Durham, absolute imposition and free individual choice need to be taken as idealized poles of a continuum, with most actual behavior falling somewhere in between, as a mixture of the two.

Having established the different forms of direct and indirect interactions between cultural and genetic replicators, we must now address certain questions regarding the *kinds* and *number* of abstract probe heads at work in cultural evolution. For example, we observed that the flow of genes through large animals is quite different from the flow through microorganisms, the former following a rigid vertical form (from one generation to another) while the latter additionally involves a horizontal exchange of genes (from one species to another, via plasmids or other vectors). In terms of the number of channels for transmission, the flow of cultural materials in human societies is quite open, and in that sense akin to the flow of genes through bacteria. Cultural replicators flow vertically in a one-to-one structure (from parents to offspring) or in a many-to-one structure (as when the adults in a community exercise pressures on a child). Cultural replicators also flow horizontally, from adult to adult (one-to-one) or from leaders to followers (one-to-many).[104]

Moreover, it may be argued that cultural evolution involves more than one searching device: while some materials replicate through *imitation* (and, hence, are analogous to birdsongs or, more generally, to memes), others replicate through *enforced repetition*: children do not simply learn to imitate the sounds and grammatical rules that make up a language, they *adopt them as a norm* or *repeat them as a rule*. (This is one minor shortcoming of Durham's analysis: he uses the term *meme* for all cultural replicators, even though some of them are transmitted as norms [e.g., his "secondary values"].) Sforza observes that linguistic norms (except for individual words) do not easily replicate across different cultures but travel along with the bodies that serve as their organic substratum. (Hence the tight correspondences he finds between linguistic and genetic maps.) He attributes this conservative tendency to the first two (vertical) mechanisms of cultural transmission.[105] The flow through horizontal channels, on the other hand, does involve imitation and so may be considered a flow of memes.

A different process is involved when the transmission involves not formalized knowledge but embodied know-how. In this case, the information in question cannot travel by itself (through books, for example) but needs human bodies as its vehicle. This kind of transmission may be compared to that involved in epidemic contagion. Braudel argues, for example, that the printing press and mobile artillery did not create a permanent imbalance in the distribution of power in Europe because they spread too rapidly across the Continent, thanks to the mobility of their practitioners. Printers and mercenaries in the sixteenth and seventeenth centuries migrated continuously, taking their skills and know-how wherever they went, spreading them like an epidemic.[106]

In thinking through the mechanisms of cultural evolution, we must take into consideration the *kinds* of entities that may be said to evolve in a given society. When studying societies that lack diversified politico-economic institutions, we may view cultural transmission in terms of replication of the whole set of values and norms which binds a particular society together. But in urban societies, institutions may also reproduce themselves with variation *individually*. The economists Richard Nelson and Sidney Winter, for instance, espouse an evolutionary theory of economics based on the idea that once the internal operations of an organization have become routinized, the routines themselves constitute a kind of "organizational memory."[107] For example, when an economic institution (e.g., a bank) opens a branch in a foreign city, it sends a portion of its staff to recruit and train new people; in this way, it transmits its internal routines to the new branch. Thus, institutions may be said to transmit information vertically to their "offspring." On the other hand, since many innovations spread through the economy by imitation, institutions may also affect each other in a manner analogous to infectious contagion.

Here we have been exploring exclusively the interactions between culture and genetics, but nonetheless we must never lose sight of the fact that the flow of replicators (whether genes, memes, norms, or routines) constitutes only half the story. The flow of matter and energy through a system (which often means the flow of biomass, either living or fossil) is of equal importance, particularly during intensifications. The role of genetic and cultural replicators (or, more accurately, of the phenotypic effects of those replicators) is to act as catalysts that facilitate or inhibit the self-organizing processes made possible by intense matter-energy flows. It is these flows that determine the nature of the thermodynamic stable states available to a system; the catalysts act merely as control mechanisms, choosing one stable state over another. Another feature of catalytic action is that low expenditures of energy can bring about high-energy transformations. An enzyme, for example, may bring about a large accumulation of a given substance by accelerating a particular chemical reaction, without itself being changed in the process (i.e., without itself participating in the larger energy transfers).

Cultural replicators may be viewed as having phenotypic effects similar to catalysis. A command given by someone of high rank in a hierarchy, for example, can set off disproportionately large flows of energy, as in the case of a declaration of war. However, the military order itself is power-less unless backed up by a chain of command that has been kept in working order through constant drill and discipline (including physical punishment for noncompliance), all of which involves enormous expenditures of bodily energy. The history of Western society in the last few centuries evidences an increasing dependency on disciplinary force to secure obedience. Therefore, we cannot be content with a description of society expressed exclusively in terms of replicators and their catalytic effects, but must always include the material and energetic processes that define the possible stable states available to a given social dynamic.

Biological History: 1700–2000 A.D.

Population explosions tend to be cyclical, like a gigantic breathing rhythm in which the amount of human flesh concentrated in one place rises and falls. These rhythms are partly the product of intensifications in food (or other energy) production, which are typically followed by depletions. The innumerable new mouths generated in the cycle's upswing

eventually eat the agricultural surpluses created by previous generations and plunge the
population into a downswing. Toward the
middle of the eighteenth century, Europe was
emerging from a cyclical downswing, a hundred years of stagnation or, at best, very slow
population growth.

Around 1750, however, several factors conspired to increase this mass of human bodies
again. A changing relationship with microbes
was beginning to transform large cities from
death traps into net producers of people. New
agricultural methods were beginning to make
intensified food production somewhat more
sustainable. And, perhaps more importantly,
massive emigration had added an escape
hatch to the dynamical system, a means to
export hungry mouths overseas, preventing
them from dragging the system into decline.

Moreover, the exportation of excess population allowed Europe to transform vast regions
of the world into its supply zones. Normally,
locally available reservoirs of biomass impose
a ceiling on population growth (technically
known as "carrying capacity"), but colonization allowed European urban centers to sur-
mount local limitations and to continue their

expansion. Europeans migrated overseas in large—eventually enormous—numbers, and they brought with them other, nonhuman "replicators": their extended families of domesticated animals and plants. Creatures not yet submitted to human control used the Europeans as vehicles for a great migration of weeds. Finally, institutional organizations also migrated, exporting their routines across the oceans to create variant replicas of themselves. Here we will first explore some of the consequences that this complex mixture had on the lands that received the migratory flow, specifically the great organic and institutional homogenizations that it effected, and then we will further address the migration's effects on the cities of Europe.

Before 1800, Europe had only sent between two and three million people to her new transatlantic colonies ("only" in comparison to the six million Africans who had been forced to migrate there). But between 1800 and 1960, sixty-one million Europeans moved across the Atlantic. Of these, the majority left for the New World in a period of seventy years. In the words of the historian Alfred Crosby:

And so the Europeans came between the 1840's and World War I, the greatest wave of humanity ever to cross oceans and probably the greatest that ever will cross oceans. This Caucasian tsunami began with the starving Irish and the ambitious Germans and with the British, who never reached peaks of emigration as high as some other nationalities, but who have an inextinguishable yearning to leave home. The Scandinavians joined the exodus next, and then toward the end of the century, the southern and eastern European peasantry. Italians, Poles, Spaniards, Portuguese, Hungarians, Greeks, Serbs, Czechs, Slovaks, Ashkenazic Jews—for the first time in possession of knowledge of the opportunities overseas and, via railroad and steamship, of the means to leave a life of ancient poverty behind—poured through the ports of Europe and across the seams of Pangaea.[108]

Pangaea is the scientific name for the hypothetical landmass the continents of the Northern and Southern Hemispheres formed when they were still joined together, many millions of years ago. New animal and plant species emerge when their reproductive communities become isolated from one another; thus the ancient breakup of Pangaea (and the consequent separation of reproductive communities) triggered an intense period of organic heterogenization. The world that witnessed the great migratory flow of the 1800s, however, was already becoming rehomogenized. As Crosby puts it, Pangaea was being stitched together again via transoceanic communications.[109] Before the 1500s, the Islamic peoples were largely responsible for the transfer of species across ecological boundaries (citrus, rice, cotton, sugarcane), but from 1500 on, the Europeans would be the main dispersants.

In five separate regions of the globe—the temperate regions of the United States, Canada, Argentina, Australia, and New Zealand—the process of rehomogenization reached its peak of intensity. These regions became, in fact, replicas of the European urban and rural ecosystems. Crosby argues that, in order for European cities to replicate themselves, to give birth to daughter cities such as Boston, Quebec, Buenos Aires, or Sydney, a whole array of species (humans and their domesticates) had to migrate together, had to colonize the new land as a team. The end result is that the temperate areas of these five countries became what he calls "Neo-Europes."[110]

There were, of course, important colonial cities outside the regions with strictly "European" climates. However, these other colonial urban centers did not reproduce the same "social ecosystem" as in urban Europe; instead, the relations between town and countryside were more like those of feudal Europe. Additionally, the neo-Europes, unlike Mexico or Peru, where the conquerors mixed with the locals, were a classic case of replacement of one gene pool by another. Finally, the tens of millions of Europeans who migrated overseas beginning in 1800 were received principally by the urban centers of the neo-Europes. These masses were not only pushed out by the population explosion at home, but also pulled in by the prospect of moving to an almost exact replica of the urban ecosystem they were to leave behind. (Having relatives abroad, the so-called stock effect, was a further pull factor.)[111]

The reason it was necessary for a whole team of colonizers to migrate across the oceans is relatively easy to grasp in the case of humans and their domesticated crops and livestock. For an urban ecosystem to work, food chains must be shortened and certain organisms must be used to redirect the flow of biomass toward the top of the hierarchy. But in addition to these domesticated species, the European migrants inadvertently imported "weeds," in this case plants with opportunistic reproductive strategies, which allowed them to colonize simplified ecosystems. Unlike many plants that thrived in the new lands only with direct human intervention, European weeds (thistles, plantain, white clover, nettles) propagated on their own, winning their own "battles" against local rivals and furnishing a key component of the food web as fodder for cattle:

> The Old World quadrupeds, when transported to America, Australia and New Zealand, stripped away the local grasses and forbs, and these, which in most cases had been subjected to only light grazing before, were often slow to recover. In the mean time, the Old World weeds, particularly those from Europe and nearby parts of Asia and Africa, swept in and occupied the bare ground. They were tolerant of open sunlight, bare soil, and close cropping and of being constantly trod upon, and they possessed a number of means of propagation and spread. For instance, often their seeds were equipped with hooks to catch on the hides of passing livestock or were tough enough to survive the trip through their stomachs to be deposited somewhere down the path. When the livestock returned for a meal the next season, it was there. When the stockman went out in search of his stock, they were there, too, and healthy.[112]

European forage grasses, which had coevolved with cattle, won their own colonization war against many local weeds, which were defenseless against the novel selection pressures (such as intense grazing) brought on by the European migration. Only in areas where large local herbivores thrived (e.g., the American Great Plains with its herds of buffalo) did the local grasses have a fighting chance.[113] In several of the neo-Europes, the weed "colonization front" raced ahead of the human wave, as if preparing the ground for it. Indeed, considering that the human colonizers were repeating past mistakes by overintensifying their exploitation of the new land (via careless deforestation, for instance), weeds played another key role, that of restabilizing the exposed soil and preventing erosion. "The weeds, like skin transplants placed over broad areas of abraded and burned flesh, aided in healing the raw wounds that the invaders tore in the earth."[114]

Weeds were not the only organic entities to spread without conscious human effort. Some plants that had been domesticated and even urbanized acquired "weedy" behavior and began winning their own propagation battles. Such was the case, for example, with peach and orange trees.[115] Even some animals (pigs, cattle, horses, and dogs) escaped human genetic control and became feral again, multiplying exponentially. These animals lost some of the qualities that domestication had imposed on them and reacquired some of the "repressed" traits of their ancestors. They, too, began colonizing the land. In Australia, pigs became razorbacks, "longed-legged and long-snouted, slab-sided, narrow-backed, fast and vicious, and equipped with long, sharp tusks."[116] In Argentina, cattle became feral, propagating in such large numbers that they stymied the growth of human populations. Here and elsewhere, these bovine multitudes formed "a cattle frontier [that] preceded the European farmers as they moved west from the Atlantic."[117]

These independent colonizers tilted the balance in the exchange of species between Europe and the rest of the world. While some American plants, including maize and potatoes, tomatoes and chili peppers, did "invade" Europe, they did so exclusively in the hands of humans, not on their own. The other spontaneous exchanges, such as the exchange of microorganisms, were also asymmetrical, despite the "gift" of syphilis which America may have bestowed on her colonial masters.[118] And, of course, the exchanges at the top of the food pyramid were heavily one-sided. Despite the influx of millions of Africans brought in by the slave trade and the masses of Asians who went overseas as indentured workers after slavery was abolished in the mid nineteenth century, by the twentieth, European migrants accounted for as much as 80 percent of the total migratory flow.[119]

Europeans benefited from this massive transfer of people in several ways. Not only did migration serve as an escape hatch from the population explosion at home, but these masses were what gave staying power to Europe's colonial ventures. Additionally, the migrants who settled in the neo-Europes achieved unprecedented fertility rates. Between 1750 and 1930, their population increased by a factor of 14, while the population of the rest of the world increased by a factor of 2.5.[120] Nonwhites were not so lucky. Slavery, which broke up families, tilted the gender ratio of populations toward males, and forced people to live in subhuman conditions, made propagation of African genes abroad very difficult.[121] Before 1800, African migrants outnumbered Europeans three to one, but their growth rates in America were vastly different: the six million slaves remained almost constant in number, while the roughly two million Europeans sextupled their population.

Part of the enormous population boom in the neo-Europes was due to the extreme fertility of their lands, in terms of both soil nutrients available after deforestation and photosynthetic potential (i.e., the amount of solar energy available for transformation into sugars; the tropics have plenty of light, but haziness and unvarying day length throughout the year make it less useful for grain cultivation).[122] Today, the neo-Europes feed the rest of the world. Even while not leading in absolute food productivity, they are the regions with the greatest food surpluses. It is no wonder that long before these colonies gained their independence they were a crucial supply region for European cities. On the other hand, the Old World had to work hard to create this reservoir for itself:

> If the discovery of America brought Europe little return in the short run, this was because the new continent was only partly apprehended and settled by the white man. Europe had patiently to reconstruct America in her own image before it began to correspond to her own wishes. Such a labor of reconstruction was not of course accomplished overnight: in the early days, Europe indeed seemed insignificant and impotent faced with the superhuman task ahead and as yet only imperfectly perceived. In fact Europe took centuries to build a world in her own image across the Atlantic, and then only with immense variations and distortions, and after overcoming a long series of obstacles one after another.[123]

Creating ecological replicas of Europe was only part of this enormous task. The European population of institutions—the whole spectrum of governmental, commercial, ecclesiastic, and educational organizations— also had to be replicated on the other side of the ocean. Europe's institutions were a complex mixture of markets, antimarkets, and rationalized bureaucracies, and their replicas across the Atlantic were equally varied. Moreover, the transformation of the American continent into a supply region involved interactions between institutions of different eras, more specifically, a mixture of different strategies for the extraction of surpluses, some ancient, some new, in a process akin to Europe's earlier self-colonization.

As urban Europe began to transform Poland and other eastern regions into a supply zone, the most "advanced" sectors of this population of institutions (the bankers and wholesalers of Amsterdam, for example) acted in collusion with the most "backward" ones, the eastern European feudal lords, to transform the free peasantry into serfs again.[124] The "second serfdom" was not a step down the ladder of progress, but rather a lateral move to a stable state (a stable surplus-extraction strategy) that had been latent in (or, available to) the dynamical system all the time. Similarly, antimarkets found entry into the American colonies through the great sugar plantations, all of which used slave labor. It was this institutional mixture that unleashed the great flows of sugar, one of the most influential forms of biomass of the colonial age.

In 1650, sugar was a luxury and its consumption a marker of status, but by the nineteenth century British industrial and agricultural workers had "sugar pumped into every crevice of their diets."[125] Sucrose made it possible to increase the caloric intake of the underclasses in a relatively inexpensive way, compared with meat, fish, or dairy products. Although it was not the only foodstuff provided by the new supply zones, it was the most efficient one in terms of converting solar energy into calories. (One acre of land produced roughly eight million calories.[126]) In this sense, sugar was at least as influential as maize or potatoes, the miracle crops Europe adopted from the New World. Large-scale sugar production also required a specific institutional mix, as sugar processing and refining demanded large amounts of capital and, hence, antimarket organizations. Sugar also generated intense profits, most of which accumulated not on the plantations themselves but in the European cities that marketed the product and provided the credit for the enterprise.[127] Sugar profits fired the European economy and later played an important role in sustaining the Industrial Revolution.

European colonization transformed the New World, and the New World in turn contributed to a transformation under way in Europe. There, the national capitals, metropolises, regional capitals, and even small towns began in the eighteenth century to escape from the biological regime of famines and epidemics to which they had been subjected since birth. Access to overseas supplies, the spread of the miracle crops, and better soil management techniques all contributed to the abatement of global famines; better transportation and communications allowed emergency aid to relieve local famines quickly. The relationship between urban masses and the microorganisms that fed on them was also changing. New epidemic outbreaks acted as catalysts for government action, and urban centers slowly began to develop new approaches to public sanitation (particularly sewage and water control) and to embrace the new technology of vaccination; that is, they slowly rejected spontaneous adaptation to disease in favor of compulsory immunization. Although deliberate inoculation had been practiced as a folk remedy since ancient times (in Turkey, for example), modern Europeans were the first to practice inoculation on a massive scale.[128] (Inoculation refers to the practice of introducing the germs that cause human diseases into the organism; vaccination, on the other hand, involves the introduction of closely related nonhuman diseases.)

Large cities were the first places to develop an unplanned accommodation with their microparasites via endemicity. This may explain why "folk" inoculation techniques first took hold, in England, in villages and small towns (where the critical human mass to sustain the stable state of endemicity did not exist), beginning with inoculations against smallpox in 1721. This does not mean, of course, that urban inhabitants were never inoculated (the elites, including the royal scions of England, were) but that, as McNeill puts it, the practice of deliberately introducing smallpox in the organism did not "take" in London and other large centers.[129] True vaccination for smallpox (using the weaker cowpox germ) was introduced in 1798 by Edward Jenner, an English country doctor, and spread from the bottom of Central Place hierarchies upward. In continental Europe, organized resistance to this practice lasted longer, and it would take the death of a king (Louis XV) to catalyze the mainland cities into action. Unlike in Britain, however, here the practice of vaccination spread from the top down: the first campaigns of vaccination took place among the elites, then the armies (by command from the top), and, finally, the civilian population.[130] In the colonies, which lacked the critical human mass and constant contact with the old world epidemiological laboratories necessary to achieve endemicity (and where, therefore, adult vulnerability to disease was greater), urban adoption of the new techniques was much swifter.

Reliable sources of food and the rise of organized medicine helped European cities and their colonial daughters leave behind the old biological regime, beginning in the mid eighteenth century. But as this bifurcation to a new stable state was taking place, as urban culture slowly detached itself from the organic constraints of famines and epidemics, the *population of institutions* that inhabited European cities underwent a momentous transformation of its own.

Military, medical, educational, and judicial institutions became, in a very real sense, much more "biological" than before: their hierarchies now relied less on tradition and symbolic gestures and began to exercise power in a form increasingly tailored to the functioning of the human body. Although the human population explosion that began in the mid eighteenth century did not cause this transformation (in armies, for example, the process had started in the sixteenth century), it did help the new breed of organizations to spread among the institutional population.

The birth of the modern hospital is a good example of the institutional transformations taking place. Western doctors had since Antiquity acquired medical knowledge almost exclusively from old authoritative texts (those of Galen, for example). The emergent medical profession, in contrast, organized itself around hospitals and could for the first time break away from textual and concentrate on biological bodies.[131] Moreover, this epistemological break did not precede the creation of hospitals, but rather was precipitated by it. The new hospitals embodied a new and different use of space, one that allowed close observation of disease and isolation of its cause. Since ocean trade routes were channels where merchandise, money, ideas, and germs all flowed together, naval hospitals provided the perfect milieu for disentangling the complex combination of factors that caused epidemics:

> A port, and a military port is—with its circulation of goods, men signed up willingly or by force, sailors embarking and disembarking, diseases and epidemics—a place of desertion, smuggling, contagion: it is a crossroads for dangerous mixtures, a meeting-place for forbidden circulations. The naval hospital must therefore treat, but in order to do this it must be a filter, a mechanism that pins down and partitions; it must provide a hold over this whole mobile, swarming mass, by dissipating the confusion of illegality and evil. The medical supervision of diseases and contagions is inseparable from a whole series of other controls: the military control over deserters, fiscal control over commodities, administrative control over remedies, rations, disappearances, cures, deaths, simulations. Hence the need to distribute and partition off space in a rigorous manner.[132]

Not only hospitals but a whole segment of the population of institutions changed during the eighteenth century. The change may nevertheless be usefully described in medical terms. Foucault pithily characterized the guiding principle behind this institutional transformation in the phrase: "treat 'lepers' as 'plague victims'."[133] In Europe, people suffering from leprosy (Hansen's disease) had traditionally been dealt with by confining them to special buildings (leprosaria) usually built outside the walls of medieval towns. There were about nineteen thousand such leprosaria by the thirteenth century.[134] The people of a plague stricken town, on the

other hand, were handled in a very different way, at least in the Mediterranean nations that had established quarantine regulations as early as
the fifteenth century. Rather than being removed from society and lumped
together in one isolated place out of sight, they were instead pinned to
their residences and observed carefully day after day by special health
inspectors, who registered their condition in writing, creating a flow of
reports linking the observers to a central command. Hence, these two
infectious diseases elicited different institutional responses, and the
insights gleaned from one could be combined with those arising from the
other — and applied to nonmedical problems. "The leper and his separation; the plague and its segmentations. The first is marked; the second
analysed and distributed. . . . Two ways of exercising power over men, of
controlling their relations, of separating out their dangerous mixtures."[135]

According to Foucault, the three elements enumerated above — systematic spatial partitioning, ceaseless inspection, and permanent registration — which had been put to work in the open space of the town, were now
combined in a novel way and applied to the closed space of the hospital.
Eighteenth-century hospitals became optical machines, places where
the penetrating clinical gaze could be trained and developed, as well as
writing machines, "great laboratories for scriptuary and documentary
methods,"[136] where every detail about visits, checkups, dosages or prescriptions, was carefully recorded. To this extent, these modern "leprosaria" had indeed internalized the quarantined urban center. On the other
hand, by administering tests and examinations on the basis of which individuals were *compulsorily assigned* to certain categories (healthy/sick,
normal/abnormal), hospitals were adapting the strategy of binary division
and branding that had been used in "treating" lepers. In short, the disciplinary approaches to disease control did not represent an advanced
"stage" in the evolution of power; rather, they were new elements added
to a mixture of materials that had been accumulating for centuries.

Nevertheless, what distinguishes the seventeenth and eighteenth centuries in this regard is the "epidemic" spread of the plague approach to
control. Before this strategy became mineralized in the form of hospitals,
it existed as a dispersed set of tactical contingency plans, heuristic
recipes, and more or less rationalized policies, with which countries bor-
dering the Mediterranean attempted to cope with the threat of biological
contagion. The formal policies had spread widely in the south, but were
unable to penetrate the towns of the northern regions because a different theory of epidemics had become "endemic" there. Medical professionals in these cities believed that "miasmas," nonorganic emanations
from decomposing organic matter, caused infectious disease, not germs

passing from one body to the next. Against this noxious, putrid air, they thought, the methods of urban quarantine were useless, and they blocked all efforts to implement quarantine policies until about 1880. In that year with the aid of a much improved microscope, scientists soon established the existence of invisible microorganisms. The miasma theory became extinct and quarantine methods soon penetrated all the cities of Europe and her colonies, and even some Islamic towns.[137]

This is only half of the story, however. As Foucault reminds us, in addition to formalized and routinized policies that may be transferred as a whole from one organization to another of the same kind, there are also methods and procedures that may diffuse individually through different types of organizations: informal techniques of notation and registration; heuristic methods for creating, correlating, storing, and retrieving files; routines for comparing documents from different fields to create categories and determine averages; techniques for the use of partitions to organize space; and methods to conduct inspections on and supervise the behavior of the human bodies distributed in that space. Thus, even though the spread of formalized policies from the Mediterranean to the north was effectively blocked by the miasma theory, this informal component could still spread contagiously, from one institutional host to the next, including nonmedical institutions. As new architectural designs for all these institutions and new examination and documentation techniques were developed, the "lepers" (students, workers, prisoners, soldiers) were indeed treated as plague victims: carefully assigned to their places, their behavior (and misbehavior) systematically watched and recorded in writing. This is not to imply, however, that medical institutions were the sole source of these disciplinary innovations. Armies were also great innovators in this area, as were some educational organizations. Foucault examines the hypothesis that these informal techniques may have spontaneously come together and interlocked to form a self-organized meshwork, or an "anonymous strategy" of domination. In his words, what formed this strategy was

a multiplicity of often minor processes, of different origin and scattered location, which overlap, repeat, or imitate one another, support one another, distinguish themselves from one another according to their domain of application, converge and gradually produce the blueprint of a general method. They were at work in secondary education at a very early date, later in primary schools; they slowly invested the space of the hospital; and, in a few decades, they restructured the military organization. They sometimes circulated very rapidly from one point to another (between the army and the technical schools or secondary schools), sometimes slowly and discreetly (the insidious militarization of the large workshops). On almost every occasion, they were adopted in response to particular needs: an industrial innovation, a renewed outbreak of certain epidemic diseases, the invention of the rifle or the victories of Prussia.... Small acts of cunning endowed with a great power of diffusion, subtle arrangements, apparently innocent, but profoundly suspicious, mechanisms that obeyed economies too shameful to be acknowledged, or pursued petty forms of coercion.[138]

In addition to entangling human bodies in a net of writing and observation, some of these institutions (mostly armies, but also schools) captured the energy of these bodies through the use of continuous physical exercises, both for training and punishment, and a system of commands based on signals that triggered instant obedience. Together, all these elements produced great "economies of scale." In the Dutch armies of the sixteenth century, for instance, the operation of loading and firing a weapon was first analyzed into its microcomponents (forty-two separate actions, each associated with a specific command), then "reassembled" in a way that reduced wasteful movements and improved coordination. An army of soldiers who had "memorized" these efficient sequences in their bodies by means of continuous drilling became more than the sum of its parts: an officer's command could trigger a synchronized series of actions (a large number of weapons firing simultaneously) producing a "solid" wall of metallic projectiles, which had a greater impact on enemy lines than random shooting.[139] Collectively, thanks to this disciplinary technique, these soldiers had now increased their power, but individually they had completely lost control over their actions in the battlefield. "Discipline increases the forces of the body (in economic terms of utility) and diminishes these same forces (in political terms of obedience)."[140]

Unlike slavery or serfdom, wherein the body is appropriated as an undifferentiated whole, here the microfeatures of bodily actions were what mattered. The new goal was to study bodies and break down their actions into basic traits, and then to empty them of their know-how and reprogram them with fixed routines. The resulting increase in the "productivity" of soldiers explains why Dutch armies were so successful in the battlefield. Although drill and discipline did not replace the older and cruder approaches (slavery, serfdom) but simply became a new addition to the growing reservoir of ways of harnessing the power of the human body, their spread nevertheless took on epidemic proportions due to the economies of scale they generated:

The style of army organization that came into being in Holland at the close of the sixteenth century ... spread ... to Sweden and the Germanies, to France and England, and even to Spain before the seventeenth century had come to a close. During the eighteenth century, the contagion attained a far greater range: transforming Russia under Peter the Great with near revolutionary force; infiltrating the New World and India as a byproduct of a global struggle for overseas empire in which France and Great Britain were the protagonists; and infecting even such culturally alien polity as that of the Ottoman empire.[141]

Thus far we have described two lines of biological history. On one hand, the eighteenth century saw Europe digesting the world, transforming it into a supply zone for the provision of energy and raw materials, a process that, at least in the case of the neo-Europes, involved a great ecological and cultural homogenization. On the other hand, European nation-states began digesting their minorities, in the sense that the new disciplinary institutions embodied homogenizing criteria of normality to which everyone was now made to conform. Much as standard English or French were normative criteria emanating from capital cities and imposed on linguistic minorities elsewhere (Welsh, Scottish, Irish; Languedoc, Catalan, Provençal), so the tests administered by various institutions to determine military performance or health status failed to reflect the cultural diversity encompassed within the borders of nation-states such as France and England.

As population growth intensified in Europe after 1750, the new masses began to be "processed" through the examining, registering, and partitioning machines that hospitals, factories, schools, and other institutions had become. These institutions acted as sorting devices, weeding out certain individuals from the reservoir of "normal" citizens who were used to fill hierarchical structures with internally homogeneous ranks. Simultaneously, surplus masses were being exported with unprecedented intensity to those temperate areas of the world where replicas of urban and rural Europe — up to the last weedy detail — had been created. In those ecologically homogenized regions, similar institutions proceeded to examine, document, and discipline the migrating human masses.

We must not, however, lose sight of the fact that just as the creation of the neo-Europes involved not only humans but also crops and livestock, so the new disciplinary institutions processed more than human bodies: animals and plants, too, fell under a net of writing and observation. Examining this other half of our biological history, its nonhuman half, will allow us to explore the role that economic institutions played in the process of organic homogenization. In particular, big business's entry into agriculture provided the impetus to apply disciplinary techniques to the members of the extended human family. Antimarkets had been involved in the flow of biomass to some extent ever since cities such as Venice and Amsterdam switched to external suppliers for their food and dedicated their own land to a variety of specialized cash crops, including oil, wine, mulberries, hemp, and flax. Traditionally wealthy merchants had purchased land as a passport to nobility; in contrast, the infiltration of the soil by antimarkets was an economic investment, and so brought with it the kind of rationalization that yields economies of scale.[142] But not until the seventeenth and eighteenth centuries did antimarket institutions' involvement in agriculture intensify, eventually to the extent that it sought to control not only flesh but genes.

Apart from the sugar imported from colonial plantations, the flow of biomass that fed the exploding population of nineteenth-century England came from regions of her countryside that had undergone an "agrarian revolution" between 1650 and 1800. An important component of this revolution was the development of new techniques for breeding livestock. The genes of farm animals had been under human control for a long time, of course, managing to escape only under rare circumstances (when domesticates became feral). But a more systematic (if prescientific) attempt at manipulating the flow of genes through generations didn't come until the agrarian revolution. The Dutch bred much larger cattle while the British bred sheep that produced superior wool, and as these breeding practices spread, so did the use of continuous observation and registration, which alone made more precise genetic control, and the consequent (sometimes damaging) genetic homogenization, possible:

> At the time of the Industrial and Agrarian Revolutions both pedigrees and economic data were recorded. Official centralized records of pedigrees were introduced with the founding of the General Stud Book in 1791 and Coates' Herd Book in 1882. Many of the genetic advantages and limitations of pedigree records are obvious. The most serious limitation has been the gradual build-up of a pedigree mystique, i.e. that pedigree animals are "superior," "prepotent" etc. by virtue of their pedigree. This has led many breeders to concentrate on the reproduction of a stereotype — the extreme of which can be seen in a number of modern dog breeds where the condition has often resulted in the incidence at high frequency of undesirable genes.... [Some pedigree monopolies and regulatory acts] certainly improved the lower level of non-pedigree English cattle by eliminating casual mating with "fringe" bulls of often inferior quality. However, such licensing acts have tended to

become too rigid in application, facilitating the "fossilization" of certain breeds in the face of changing economic requirements.[143]

Historically, pedigree breeds have always tended to become hierarchical structures, wherein a small, dominant group of breeders supplies genes to subordinate ranks, called "multipliers," which in turn pass them on to yet lower ranks, in a completely top-down gene flow. This tightly constrained flow was supposed to guarantee uniformity and superior quality, and yet there is evidence that bottom-up flow can, in some circumstances, produce breeds superior to homogenized pedigrees.[144] At first, however, the pedigree breeds' productivity was great, and this allowed the hierarchical pedigrees that emerged in eighteenth-century England (especially sheep and pigs) to thrive and then spread, aided by large agricultural shows where new machinery and champion breeds were exhibited.[145] Thus, much as transoceanic navigation had accelerated the genetic homogenization of certain parts of the world (by allowing massive transfers of species), the creation of monopolies and oligopolies around the flow of livestock genes fostered the destruction of genetic heterogeneity in Europe.

These genetically "well-disciplined" animals were only one component of the agrarian revolution. There were new crops as well, particularly fodder crops, and a few new machines (the seed drill, for example), but the most important innovation was the introduction of more routinized methods for the production of food, for both humans and livestock. And, of course, typical of any endeavor of antimarket institutions, these methods were implemented on a large scale. The new synergistic combination of elements was called "the Norfolk system," after the region in England where it first triumphed. We must distinguish, however, two different components of this system. Unlike the case of large-scale management and labor discipline, the basic meshwork that gave the new system its self-sustainability was not introduced by big business but was the creation of market economies. The dynamic cities of fifteenth-century Flanders (Bruges, Ypres, Ghent) stimulated their countrysides into producing the basic innovations. In Flanders, as one eminent historian has put it, urban life spread like "an infection which roused the peasant from his age-long torpor."[146]

At the time of the Norfolk system's creation—that is, before it was adopted by antimarket institutions and before it was called the Norfolk system—the most widespread system of agriculture consisted of simple crop rotation: a farm was divided into two (or more) parts, one used for grain crops and the other left fallow, not to let the soil "rest" (soils do not spontaneously recover their fertility in a single season), but to allow "farmers to keep weeds at bay by interrupting their natural life cycle with the plow."[147] Denying soil nutrients to weeds and keeping predators from eating livestock were the primary ways in which humans shortened food chains; consequently, crop rotation was a crucial component of the old method. The Flemish contribution to the agricultural intensification was to eliminate the fallow period by alternating grain crops with fodder crops (such as clover). As the Dutch historian Jan De Vries has argued, population growth often trapped the old method into a vicious circle: as demand for human food increased, more land was devoted to grain production and less to pasture, which diminished herd sizes as well as the amount of manure available, and this in turn reduced soil fertility. As yields declined, a higher percentage of the land had to be used for grain, exacerbating the overall decline.[148]

Turning this vicious circle into a virtuous circle involved reorganizing the rotation system so that arable lands could contribute to the fodder supply. This meant planting clover (or, later on, alfalfa or turnips) instead of letting land lay fallow. Feeding these crops to cattle, in turn, allowed herds to increase in size and hence to multiply manure supplies. Moreover, continuously feeding manure back into the soil, as well as using fodder crops to bind the soil and prevent it from escaping the system via water or wind erosion, meant *tightening the nutrient cycles*, a process that takes place spontaneously in mature ecosystems and greatly contributes to their resiliency.

Flanders, a highly urbanized area, was among the least feudalized regions in Europe, which goes a long way in explaining why the new agricultural methods developed there. That the region was not feudal, however, does not mean it was "capitalist." As I have repeatedly pointed out, private property and commercialization do not necessarily imply the presence of antimarkets. Indeed, De Vries explicitly marks this difference by developing two separate models to analyze the evolution of this agricultural regime, one based on market involvement, the other antimarket.[149]

The Flemish method, further developed in the Netherlands, soon found its way to England, where it was employed on a large scale and subjected to disciplinary management. Only after the English modified the system was there a truly "capitalist" agriculture. In eighteenth-century England, vast tracts of land were submitted to the new intensive methods and enclosed on all sides with hedges. Landowners and the farmers of large holdings reaped the benefits of the new productivity, while countryside strata (landlords, tenants, and de-skilled laborers) hardened, reducing the number of intermediate classes (small holders, rural tradesmen).[150]

These "well disciplined" lands fed the growing British population, a substantial portion of which would provide the raw muscular energy and skills for the new industrial towns and conurbations for two centuries.

By the mid 1800s, large-scale agriculture in England was eclipsed by similar but even larger enterprises in the neo-Europes: the United States, Australia, and Argentina. In these places (as well as in Siberia) the meshwork that characterized the Norfolk system acquired new nodes (in the form of new machines, such as McCormick's reaper, which automated some aspects of harvesting) and much greater proportions.[151] Moreover, the very tight nutrient cycles that characterized the Norfolk system were suddenly split wide open as natural and artificial fertilizers began to be used in agricultural production.[152] In the United States, for example, fertilizer began to flow in from as far away as Chile.[153] Not only were the nutrient cycles opened to inputs from distant origins, their outputs were also divorced from the soil: the nitrogen and phosphorous in many fertilizers were not completely absorbed by plants (almost half of these nutrients was wasted) and escaped the Norfolk system, seeping into the groundwater and overenriching it in a process called *eutrophication*.[154] Moreover, every nutrient flow that came from outside the farm was one more point of entry for antimarkets, and, hence, represented a further loss of control by the food producers. A century later, as we will see, corporations would genetically engineer crops that required excessive fertilization, thus etching entry points for antimarkets into the crops' very genes.

Although this kind of near-total genetic control over the flow of plant biomass would not be realized until the late twentieth century, the disciplining of plant genes was already practiced in the late nineteenth and early twentieth centuries. Plant pedigree hierarchies lagged behind their livestock counterparts, but when they finally materialized the degree of human control over them was much greater. And that manipulation of plant genes would lead to a process of genetic homogenization that dwarfed all earlier homogenizing trends.

As often the case, more than one kind of institution was involved in this process. In particular, certain government agencies in the neo-Europes led the way to the creation of plant pedigrees. In 1862, as the western frontier was officially opened in the United States, a department of agriculture (the USDA) was created for the purpose of collecting, propagating, and distributing seeds for crop plants. Land-grant universities and experimental-agriculture stations were also created to help develop better plant varieties and multiply them; that is, planting them only as a source of genetic materials.[155]

The first plant to be captured in the net of observation and writing was corn, chosen for the accessibility and manipulability of its sexual organs. By 1896, one of the agricultural stations had developed the technique of *inbred lines*: repeatedly crossing a given strain with itself, until certain genes were eliminated and others driven to fixation. Despite the "pedigree mystique," it soon became obvious that such extreme homogeneity actually had damaging effects on the plants, but by 1905 a new technique had been developed to compensate for this: crossing two different inbred lines of corn kept the "desirable" traits in their progeny while eliminating some of the undesirable ones. This process produced what came to be known as "hybrid corn":

> Although hybrid corn was first introduced to farmers in 1926, only about one percent of the acreage in the Corn Belt was planted to hybrid varieties by 1933. This changed rapidly, however, and by 1944 more than eighty-eight percent of the Corn Belt was planted to hybrid corn. Yields increased dramatically; "corn power" had arrived.... With hybrid corn, only those who knew the parent lines and breeding sequence knew how to make the high-yielding hybrids—called a "closed pedigree" in the business—and this knowledge was legally protected as a trade secret. More importantly from the business standpoint, farmers could not save and reuse hybrid seed the following year and obtain the same yield, since "hybrid vigor" would decline with continuing use of the seed. Farmers had to return to the seed companies to buy new seed each year.[156]

Hybrid corn was the product of one homogenizing operation (which created the parent inbred lines) followed by one or more heterogenizing operations (crossing the inbreds to maintain hybrid vigor). However, due to the hierarchical structure of pedigrees and of the oligopolistic practices behind their spread, the whole process was crowned with another (and more powerful) homogenization: in the nineteenth century the gene pool of American corn was rich in variety, but by World War II most of those genes had been driven out and replaced by the cloned genetic materials from a few parental lines.

At the time, this process was considered "progress," but the homogenization of the Corn Belt (and other food-producing regions) was indeed extremely dangerous. Although crops and livestock have from ancient times been as susceptible to epidemics as human populations, a certain degree of heterogeneity in their genetic makeup protected them from extinction. While some of the individual plants in a field would perish under the onslaught of disease, others would survive and continue the line. But when 80 percent of the plants in a given population are virtually clones, the moment a new microorganism hits on a "genetic window," there are no obstacles to its spread. This is exactly what happened several decades ago, when a new fungus found an entry point that enabled it to elude hybrid corn's defenses:

> Reproducing rapidly in the unusually warm and moist weather of 1970, [the fungus's] spores carried on the wind, the new disease began moving northward toward a full-scale invasion of America's vast corn empire.... The new fungus moved like wildfire through one corn field after another. In some cases it would wipe out an entire stand of corn in ten days.... The fungus moved swiftly through Georgia, Alabama, and Kentucky, and by June its airborne spores were headed straight for the nation's Corn Belt, where eighty-five percent of all American corn is grown.[157]

As it happened, after a good part of the year's crop had been destroyed, a change in the weather and emergency measures that were taken saved the day. But the epidemic had already made clear the dangers of homogenization and the long-term consequences of decisions made three or four decades before. Moreover, after the initial successes with corn, hybridization techniques spread to other plants (e.g., alfalfa and sorghum) and then, in the 1940s, to animals—first chickens and later on cattle.[158] The resulting genetic uniformity has made many industrialized nations "gene poor" countries that now view with envy the genetic resources of their "gene rich" underdeveloped neighbors.

Even before hybridization techniques had "genetically disciplined" corn, the earlier successes of livestock pedigree hierarchies had inspired some scientists to dream of applying selective breeding techniques to human beings. In the second half of the nineteenth century, when Francis Galton coined the term "eugenics," a widespread movement sought to give disciplinary institutions control over the flow of human genetic materials. The movement gained momentum in the early twentieth century, particularly after the rediscovery of Mendel's work on heredity and the establishment of genes as the carriers of hereditary information. The idea of "improving" human beings through selective breeding was not new (it is at least as old as Plato[159]), but in the early twentieth century it meshed well with the development and spread of hospitals, prisons, and other institutions that routinely partitioned, examined, and documented human beings. In other words, while the dream of "genetic hygiene" may be old, the tools for its implementation were just reaching maturity and spreading through the population of institutions. Special organizations such as the Eugenics Record Office came to life in the United States (as well as in England and Germany) and took on the task of subjecting the human gene pool to the system of continuous writing and observation:

> Researchers at or affiliated with these laboratories gathered information bearing on human heredity by examining medical records or conducting extended family studies, often relying upon field-workers to construct trait pedigrees in selected populations—say, the residents of a rural community —on the basis of interviews and the examination of genealogical records. ... By 1926, as a result of its surveys and studies, the Eugenics Record Office had accumulated about 65,000 sheets of manuscript field reports, 30,000 sheets of special traits records, 8,500 family trait schedules, and 1,900 printed genealogies, town histories, and biographies.[160]

Although the scientific value of most of these data was minimal (only the relatively few heritable traits that depended on a single gene could be trapped in this net), its social consequences were not. Informed by very primitive thinking about genetics, where even the most complex disposi- tions were reified into simple entities and linked with single genes, Ameri- can eugenicists managed to involve several institutions directly in the control of the flow of human genetic materials. Beginning with Indiana in 1907, over twenty states passed compulsory sterilization laws in an overt attempt to eliminate certain genes from the pool. Despite the fact that most of these "genes" were spurious (e.g., drunkenness, feebleminded- ness, and vagrancy "genes"), thousands of people were sterilized and continued to be forcefully disconnected from the gene pool even after the eugenics movement had died. Additionally, fearing the great influx of southern European blood, the tail end of the massive human wave that came to the neo-Europes in the nineteenth and twentieth centuries, immi- gration authorities passed laws to restrict the kind of genes that came into the United States. Although the Immigration Restriction Act of 1924 did not explicitly phrase its policy in eugenic terms, it is clear (as Stephen Jay Gould has argued) that it was intended to favor the entry of northern European "stock" at the expense of gene pools deemed inferior.[161]

The practice of immigration control is particularly relevant here because it involved a new type of examination technique that is still used today as a "sorting device": the IQ test. Originally created (by Alfred Binet between 1905 and 1908) as an aid to detect children who may need special educa- tion, it was transformed by American eugenicists into a routine device for testing and ranking all children and adults according to their (supposedly heritable) mental worth.[162] An essence of "rationality" was postulated, reified into a "thing" in the brain, and then associated with a single "gene" whose presence or absence from the gene pool was susceptible to institutional manipulation. Regardless of the fact that the test mostly measured familiarity with American culture, mastery of the "the arcana of bowling, commercial products, and film stars,"[163] it became a routinized procedure to brand immigrants according to their genetic endowment. It was also directly connected with the sterilization campaign, since low IQ scores were thought to signal "feeblemindedness," a supposedly heritable condition that endangered the integrity of the American gene pool.

Although eugenics was eventually discredited when Nazi Germany showed the world just what such genetic "improvement" could lead to if implemented on a large enough scale, this did not mean that the human body escaped the net of writing and observation into which it had been drawn two or three centuries earlier; there were other means of controlling its capabilities which were unrelated to crude genetic cleansing campaigns. We may divide these into two types, following the distinction biologists make between *soma* and *germ line*: the latter refers primarily to cells with reproductive capacity (eggs and sperm), but may also be said to include all the tissues and organs that make up our reproductive system, while the former includes all the other systems (digestive, muscular, nervous, etc.) that form the rest of the body. In terms of social control over the soma, it has principally been the male body that has suffered the effects of disciplinary techniques. Not only were drill and surveillance developed in exclusively male armies, but large masses of male bodies were used as cannon fodder from the Napoleonic Wars through World War I. (In the latter, an entire generation was used to "feed" enemy artillery.) In terms of the germ line, on the other hand, the female body has borne the brunt of intense examination and registration techniques.

A very important institutional encroachment on the germ line occurred in the United States during the nineteenth century through the ascendance of obstetrics and gynecology. Between them, these new specialties managed in a few decades to acquire a virtual monopoly over the methods and practices used to assist in childbirth. "In the [early] twentieth century, physicians pushed for 'obstetrical reform,' which largely eliminated midwives and moved birth from the home to the hospital. While in 1900, fewer than 5 percent of American women delivered in hospitals, by 1940, about half did and by 1960, almost all."[164] As medical studies (by doctors) have revealed, during the period of time in which hospitals took over from traditional practices this crucial position in the flow of genetic materials, obstetricians were causing more damage to women than midwives ever did. Aggressive use of forceps tended to result in torn birth canals, and lack of hygiene spread diseases among the infants:

> Increased physician attendance at birth did not result in improved outcome for mothers and babies. As the percentage of births attended by midwives decreased from 50 to 15 percent, perinatal infant mortality increased. During the first decade of the twentieth century, midwives in New York were significantly superior to doctors in preventing stillbirths and childbed fever. For example, Newark's maternal mortality rate of 1.7 per 1,000 from 1914 to 1916 among mothers delivered by midwives compared most favorably to the 6.5 per 1,000 rate in Boston, where midwives were banned.[165]

In the long run, as rationalization and routinization gave rise to economies of scale, hospitals may have become better places for humans to be born, at least in terms of decreased mortality. The problem, as with assembly-line factories, was that this increased "productivity" came with hidden costs in terms of loss of control (for the women giving birth). As with all disciplinary institutions, a true accounting must include those forces that increase (in economic terms of utility) and those that decrease (in political terms of obedience). Sedated women giving birth in hospitals not only lost control over decisions made during labor (for instance, whether or not any surgical intervention is required) but also over other functions later on:

> In the 1930s physicians began replacing the woman's breast milk (which an early Gerber advertisement for baby formula called "a variable excretion") with formula, a product increasingly available from drug and milk companies.... To discourage nursing on demand, they separated mother and child. They established rules requiring feedings at intervals of no less than four hours.... In the nurseries, babies were fed supplemental bottles without the mother's knowledge. Consequently, the babies were not hungry when brought to the mother. Without sufficient suckling the mother's milk dried up.... By the 1940s the proportion of women breast-feeding, with or without supplemental bottles, had dropped to 65 percent. By 1956, it was down to 37 percent; by 1966, 27 percent.[166]

Despite the current revival of midwifery (and breast feeding), the transfer of birth from private homes to public spaces of observation and writing was an institutional encroachment on the human germ line. And this takeover complemented the earlier snaring of our soma in a similar net of compulsory tests and records. The French military, which pioneered the

171

routinization of industrial production in its eighteenth-century arsenals, was perhaps the first to combine the effects of drill with those of hygiene and medicine to produce not only obedient but healthy bodies. The massive armies of urban proportions with which Napoleon conquered Europe were epidemiologically akin to cities. Only the combined effects of compulsory vaccination, a ritual attention to cleanliness, and a medical corps with a clear chain of command made possible these otherwise imprudent mixtures of recruits from regions not normally in close contact with one another.[167]

Thus far we have explored the two halves of *our* biological history, the history of our own flesh and blood as well as the nonhuman genes and biomass under our control. However, as we have already seen, the history of urban alimentary pyramids needs to be complemented by analysis of the larger biological meshwork of which cities and towns are a part. More specifically, we need to return to the microscopic component of those food webs, the world of infectious diseases that continue to feed on our bodies and hence short-circuit our tightly focused biomass flow. Moreover, microorganisms interact not only with our organic bodies but also with our institutions, exerting selection pressures on them and thereby acting as sorting devices for the routines that these institutional replicators transmit vertically and horizontally.

Much as the plague stimulated the creation of the methods and routines that would later on mineralize into hospitals, the cholera epidemics of the nineteenth and twentieth centuries catalyzed into existence a number of urban institutions concerned with public health and hygiene. In British towns, local boards of health emerged as a response to the first outbreak in 1832. A second wave hit in 1848, and this time a centralized agency was created to implement far-reaching programs of public sanitation. Cholera is a waterborne disease, and so the response to it necessarily involved new systems of water supply and sewage disposal. The intrusive character of the infrastructure that was needed (pipes running under private property, for example), as well as the then-dominant miasma theory of epidemics (which favored air and earth as transmitters), generated resistance to the project, and it took the intense fear that cholera inspired to overcome these obstacles. Similar situations cropped up in other parts of Europe, as well as in the lands Europeans had settled:

> Spread [of the new policies] to other countries occurred relatively rapidly, though not infrequently it took the same stimulus of an approaching epidemic of cholera to compel local vested interests to yield to advocates of sanitary reform. Thus, in the United States, it was not until 1866 that

a comparable Board of Health was established in New York City, modeled on the British prototype and inspired by identical apprehensions of the imminence of a new cholera epidemic. In the absence of this sort of stimulus, such a great city as Hamburg persisted in postponing costly improvements of its water supply until 1892, when a visitation of cholera proved beyond all reasonable doubt that a contaminated water supply propagated the disease.[168]

McNeill calls cholera the first "industrial disease" not because it originated in factory towns (it did not) but because it had reached Europe from India thanks to new transportation technologies such as the steamship and the railroad. These channels allowed microorganisms to travel farther and faster than ever before: a cholera epidemic that began in Bengal in 1826 reached eastern Europe in 1831, the United States in 1832, and Mexico in 1833.[169] Consequently, cholera also catalyzed the first attempts at international cooperation in responding to epidemics. (As early as 1831, Europeans were collaborating with Egyptian authorities in tracking the course of the disease.) When steamships began connecting the world's maritime gateways around 1870, the range of habitats that could be colonized not only by germs but by weeds (rats and their fleas) increased greatly. In the 1890s, a new epidemic of bubonic plague broke out in China and by 1894 had reached Canton and Hong Kong. From there steamships carried the infected rats and fleas to other ports, from which, in turn, the disease spread into burrowing rodent communities elsewhere. Although international teams of doctors and a number of prophylactic measures managed to contain the spread of plague to humans, even today new versions of plague are evolving in underground rodent "cities," some capable of infecting people:

> Plague was brought by ship to the northwest of America around 1900. About 200 deaths were recorded in the three-year San Francisco epidemic which started just after the earthquake in 1906. As a result, the western part of the U.S.A., particularly New Mexico, is now one of the two largest residual foci of plague (in mice and voles particularly) in the world — the other is in Russia. The plague bacillus has spread steadily eastwards from the west coast and in 1984 was found among animals in the mid-west. The wave front has moved on average about 35 miles a year.... If, or rather, when, plague reaches the east coast of the U.S.A. with its large urban areas, the potential for a serious epidemic will be considerable. New York, for example, has an estimated rat population of one rat per human; and mice — also effective disease carriers — probably number more.[170]

As this example illustrates, the fact that modern medicine has gained a larger measure of control over microorganisms does not mean that we have ceased to form a meshwork with bacteria, viruses, plasmodia, fungi, and other "weeds." But the command elements in the overall mixture have increased, and this has had important historical consequences. To begin with, the medical and public health institutions that were generated in our clash with epidemics managed to push cities across a threshold around the year 1900: for the first time in the millennium (and perhaps in history) large cities were able to reproduce their human populations without a constant flow of immigrants from the countryside. The city became, in a sense, self-reproductive.

Then international emigration flows received a boost as military medicine, now able to implement hygienic and immunological programs by command, allowed armies to break away from old biological regimes and opened new areas for colonization. Some of the great colonial enterprises of the late nineteenth century—the opening of the Panama Canal by the United States (in 1904) and the carving up of the African continent by several European powers—were made possible by the increased control over malaria and yellow fever achieved by military medicine. The vector of both diseases (mosquitoes) was brought into the disciplinary net by a rigorous sanitary police "supported and sustained by meticulous observation of mosquito numbers and patterns of behavior."[171]

But the real breakthrough in the attempt to submit microorganisms to pyramidal control occurred when laboratories learned how to turn microbe against microbe on an industrial scale. This took place during World War II, with the development of a series of new chemicals, such as penicillin and sulfas. When the term *antibiotic* was introduced in 1942, it was defined as any chemical substance produced by a microorganism capable of disturbing a vital link in the metabolism of another one, thus killing it or inhibiting its growth.[172] (Today some antibiotics are chemically synthesized, so the definition has been broadened.) These naturally occurring substances may be the product of arms races between microbes (similar to those between predators and their prey), and their existence had been known for several decades prior to the war. But not until the 1940s did the war on disease possess the industrial methods needed to force a "microbial proletariat" to mass-produce these chemical weapons.

Although antibiotics did prove decisive in winning the first battles, they did not allow medical institutions to win the war. The problem was that, as it turned out, microbes offered these weapons a constantly moving target. The flow of genes in microorganisms, unlike large animals and plants, is not rigidly hierarchical; even those microbes that reproduce sexually (and thus channel genes "vertically," as we do) also communicate "horizontally" with one another, freely transferring pieces of genetic information across strains and even species. Soon after World War II, genes that conferred resistance to antibiotics were promptly transferred from one species of bacterium to another. Since penicillin's initial use in 1941, a majority of its targets (staphylococci) have become resistant to it.[173] Pumping massive amounts of antibiotics into animal and human intestinal tracts worsened the situation by creating the perfect environment for the selection (on an equally massive scale) of new resistant strains. Today, nearly every disease known to medicine has become resistant to at least one antibiotic, and several are immune to more than one. It seems clear now that we will continue to form a meshwork with the microworld despite all the advances in medical science. A similar point applies to plant and insect "weeds." Because of the massive applications of DDT (and other members of its chemical family) to shorten urban food chains, some scientists believe that the only weeds that will be around in urbanized regions by the year 2000 are those resistant to these pesticides.[174]

Thus, a new arms race developed, this time between hierarchical medical institutions and the rapidly evolving meshwork of microbes. In the latest round of this contest, the very machinery behind the horizontal transfer of genes among bacteria was recruited to serve the bacteria's very enemy. The mechanism involves at least two components: jumping genes and a vector of transmission (plasmids, transposons). The discovery that pieces of genetic information can move around in a chromosome dates to the late 1940s, but it took decades before the entrenched orthodoxy could accommodate the new ideas. Today we know that genes not only can move around inside the nucleus, they can also "jump" out into the cytoplasm and become incorporated into organelles (such as plasmids), which reproduce on their own within the cell. Plasmids can travel from one cell to another (or one bacterium to another) and deliver the "jumping gene," which then incorporates itself into the nuclear DNA of the new cell and thus becomes heritable. This mechanism may explain how resistance to antibiotics spread so rapidly among the population of microbes.

With the discovery of gene-splicing and gene-gluing enzymes, as well as the other techniques of biotechnology, human researchers were able to exploit this mechanism and take genetic materials from one living creature, attach them to a plasmid (or other vector), and then inject them into a different creature, in effect, creating "chimeras": animals, plants, or microbes with the genetic characteristics of two or more different species.[175] The practical value of chimeras for the arms race between medical institutions and microbial evolution is this: genes that code for specific enzymes (or other proteins) with potential medical applications can now be incorporated into an easy-to-cultivate cell, using its own machinery to "translate" the gene into a protein. By cloning this chimeric cell repeatedly, large populations of protein producers can be created and their product harvested through a variety of methods.

Paradoxically, the very procedures employed to deny microparasites access to the urban flow of biomass allowed *macroparasites* (especially antimarket institutions) to insert themselves at multiple points in the food chain. As we saw above, this trend began with the introduction of chemical fertilizers (as well as herbicides and insecticides), which are manufactured far from the farm and which split open the nutrient cycles that had been closed for centuries. While a century and a half ago American farms produced most of what they needed (running on tight nutrient cycles), today they receive up to 70 percent of their inputs (including seed) from the outside.[176] Biotechnology is accelerating this trend, but it did not create it.

Take, for example, the green revolution of the 1950s. New plant hybrids with genes that directed most photosynthetic activity to the production of edible grain (as opposed to inedible stems) were introduced in the Third World, with the admirable goal of making those countries nutritionally self-sufficient. And, indeed, the much higher yields of these "miracle" plants did for a while strengthen the food base of countries such as Mexico, the Philippines, and India. The catch was that the new breeds required large amounts of outside inputs (fertilizer) to perform their miracles, and in the absence of chemical fertilizer their yields were not nearly as impressive. The situation was similar to that of steam power: in order to get high outputs of mechanical energy, intense inputs of coal were needed. In other words, this kind of setup profited from economies of scale and therefore benefited large farmers, triggering a process of consolidation in which many small farms disappeared. Open nutrient cycles also made farmers vulnerable to outside monopolies: when the Arab oil cartel began raising prices in the early 1970s, fertilizer costs increased dramatically and the green revolution collapsed. Worse yet, clones of the new plants now dominated the local gene pools and many genetic materials of traditional varieties (which did not depend on fertilizer) had been lost, making it very hard to turn back the clock.[177]

The homogenization of the genetic base of crops and livestock reached high peaks of intensity in the last few decades. And the genes that are being selected now, unlike during the Green Revolution, are not those that increase the nutritional value of biomass, but rather its *adaptability to homogeneous factory routines*. For instance, in the 1950s and 1960s, manufacturers of farm machinery worked together with plant breeders to fit new vegetable varieties to the demands of routinization and rationalization. Genes that caused vegetables to yield uniform shapes and sizes, as well as to mature simultaneously to allow harvesting at the same time, made it easier to adapt vegetable production to machines and to factory schedules:

> Crops in the field must first meet the tests of yield, uniform growth, and simultaneous maturity. After this, their fruit or kernels must be able to withstand the rigors of mechanical harvesting, repeated handling, and various kinds of transport from one point to another. Next come the trials of steaming, crushing, or canning. In some cases, the raw agricultural crop must "store well" or "travel well," or be good for freezing or frying. And genes are the keys to meeting each of these steps in the food-making process; the genes that control the field-to-table characteristics of every crop from broccoli to wheat. In this process the genes that matter are those of yield, tensile strength, durability, and long shelf life. However, the genes for nutrition —if considered at all—are for the most part ignored.[178]

In some cases, the genetic materials behind "well-disciplined" processing properties are in direct opposition to those improving nutritional value (that is, breeding for one eliminates the other). Consequently, the latter could very well disappear from these new plants, and as clones of the new varieties spread, the genes of old varieties will begin to disappear from the gene pool. Hence, the evolution of crops (and livestock) is truly being driven from the processing end of the food chain. A few centuries ago, cultures (Islamic, European) were the main vectors for the transmission of genes across ecosystems; today, corporations have inherited this homogenizing task. McDonalds, for instance, is now the main agent of propagation of the genes behind the Burbank potato; the Adolph Coors Company, of the genes for the Moravian III barley; and the Quaker Oats Company, of the genetic base of a few varieties of white corn hybrids.[179]

Biotechnology is bound to intensify this homogenization even more. Although most biotechnological innovations were developed by small companies, these innovators are being digested through vertical and horizontal integration and incorporated into the tissues of multinational corporations, in many cases the same ones who already own seed, fertilizer, and pesticide divisions. Rather than transferring pest-resistant genes into new crop plants, these corporations are permanently fixing dependence on chemicals into crops' genetic base. For instance, corporations such as

DuPont and Monsanto, which create weed killers, need to develop crops that withstand these chemical attacks. Thus they are transferring the genes from weeds that have developed resistance to these substances to new crop varieties, and thereby genetically freezing farmers' dependence on external inputs.[180]

Farm animals are suffering a similar fate. For instance, "well-disciplined" pigs and cows—that is, livestock that are capable of withstanding the stresses of confinement and that possess the uniform characteristics demanded by meat-packaging specifications—are today being bred or engineered. Moreover, the techniques used to exercise tighter control over the flow of genes across animal generations (artificial insemination, in vitro fertilization, and embryo transfer) were soon applied to humans, once the techniques had proven themselves "safe" and effective. Needless to say, despite the recent revival of eugenics (exemplified, for example, in the creation of human sperm banks[181]) and the ongoing human genome program (which aims for complete genetic self-knowledge by the first decade of the new millennium), the homogenizing consequences for our species will not be nearly as dramatic as for our crops and livestock. Given that our flesh does not flow in the urban food pyramid, we hardly risk being forcefully "evolved" by food processors and packagers. And yet, as we saw before, there are real dangers in human genetic manipulation, though the dangers lie elsewhere.

The institutional strategies of continuous examination and recording that had been developed to fight the plague were first applied to humans, and only later to plant and animal pedigrees. Genetic tests, such as those being developed to screen us for heritable diseases (the main rationale behind the human genome program), will be added to the growing arsenal of examination procedures already used by many institutions to screen and sort human beings. Moreover, many of the genetic diseases that will in the near future become detectable through genetic testing *lack any effective medical treatment or cure*. Under these circumstances, all a genetic test will do is brand certain individuals as carriers of the disease. Thus, as some critics of genetic testing have argued, "We risk increasing the number of people defined as unemployable, uneducable, or uninsurable. We risk creating a biological underclass."[182]

In this chapter we have followed the history of the different biological components of urban dynamics. These must be added to the flows of mineral matter-energy that traverse Western urban societies. We have noted repeatedly that, in addition to the construction materials for our homes and bodies (stone and genes, live and fossil energy), a variety of "cultural materials" flow through and accumulate within our cities. However, with some exceptions, we have used this phrase in a largely metaphorical way, to suggest that, in this case too, we are dealing with nothing but "stuff." It is time now to attempt to excise this metaphor, to explore cultural accumulations in detail and decide whether they, too, are merely sedimentations hardened by time and sculpted by history, intercalated heterogeneities connected by the local action of catalysts, replicating structures blindly exploring a space of possibilities. The following chapter focuses on language, of all the different manifestations of human culture, not only because it is the one structure that makes us unique among living creatures, but also because linguistic structures have undergone a similar process of intense homogenization, involving a variety of institutions, such as academies and schools, newspapers and news agencies. Our exploration of the routinization and uniformation of linguistic materials will reveal that an even wider segment of the population of institutions was involved in creating the homogenized world we inhabit today.

3: MEMES A

ID NORMS

Linguistic History: 1000–1700 A.D.

Human languages are defined by the sounds, words, and grammatical constructions that slowly accumulate in a given community over centuries. These cultural materials do not accumulate randomly but rather enter into systematic relationships with one another, as well as with the human beings who serve as their organic support. The "sonic matter" of a given language (the phonemes of French or English, for instance) is not only structured internally, forming a system of vowels and consonants in which a change in one element affects every other one, but also socioeco-nomically: sounds accumulate in a society fol-lowing class or caste divisions, and, together with dress and diet, form an integral part of the system of traits which differentiates social strata. A similar point can be made about lexical materials and grammatical pat-terns. As the sociolinguist William Labov has observed, a language communicates informa-tion not only about the world but also about the group-membership of its human users.[1]

This section outlines the broad history of linguistic accumulations in Europe between 1000 and 1700 A.D. and the more or less sta-ble entities they gave rise to, particularly when linguistic materials accumulated within the walls of a city or town. Thus, as the sounds, words, and constructions constitut-ing *spoken* Latin sedimented in the emerg-ing urban centers of the southern regions of Europe, they were slowly transformed into a multiplicity of dialects, certain of which eventually developed into modern French,

Spanish, Portuguese, and Italian. (And a similar process transformed the Germanic branch of Indo-European dialects into various modern tongues, including English, German, and Dutch.)

Here we will explore the idea that the different structure-generating processes that result in meshworks and hierarchies may also account for the systematicity that defines and distinguishes every language. In particular, each vowel and consonant, each semantic label and syntactic pattern, will be thought of as a *replicator*, that is, as an entity that is transmitted from parents to offspring (and to new speakers) as a norm or *social obligation*. A variety of social and group dynamics provides the selection pressures that sort out these replicators into more or less homogeneous accumulations. Then, other social processes provide the "cement" that hardens these deposits of linguistic sediment into more or less stable and structured entities. This is not, of course, a new idea. Indeed, it would seem to be the basic assumption behind several schools of historical linguistics, even if it is not articulated as such. This is particularly clear in the role that *isolation* plays in these theories. Much as reproductive isolation consolidates loose accumulations of genes into a new animal or plant species, *communicative isolation* transforms accumulations of linguistic replicators into separate entities. In the words of the evolutionary linguist M. L. Samuels:

> It is ... the mere fact of *isolation* or separation of groups that accounts for all simpler kinds of [linguistic] diversity. Complete separation, whether through migration or geographical or other barriers, may result in dialects being no longer mutually intelligible; and thus, if there is no standard language to serve as a link between them, new languages come into being. Lesser degrees of isolation result in what is known as a dialect continuum — a series of systems in which those nearest and most in contact show only slight differences, whereas the whole continuum, when considered from end to end, may show a large degree of total variation. Dialect continua are normally "horizontal" in dimension, i.e. they occupy a region in which fresh differences ... continually appear as one proceeds from one village to the next; but in large towns they may also be "vertical," i.e. the different groups belong to different social strata in the social scale.[2]

Thus, the flow of norms through generations (and across communities) may result in both meshworks and hierarchies. A continuum of dialects is a meshworklike collection of heterogeneous elements to the extent that each dialect retains its individuality and is articulated with the rest by overlapping with its immediate neighbors. It is this area of overlap — the common sounds, words, and constructions between nearby dialects — that articulates the whole without homogenization: two dialects on the outskirts of the continua may be quite different (or even mutually unintelligible), and yet they are connected to each other through intermediate dialects. For instance, the dialect of medieval Paris (now referred to as "Francien") was connected to the dominant dialect of Italy (Tuscan) by many intermediate forms: a whole set of French, Franco-Provençal, and Gallo-Italian dialects. (Rather sharp transitions, or *isoglosses*, do occur in this continuum.)[3]

Conversely, the dominant variants of the language of a given city, as well as dialects that have become "standard" (such as *written* Latin in the Middle Ages), are relatively homogeneous entities, in which the norms have been fixed either through the deliberate
intervention of an institution (in the case of "standards") or by the "peer pressure" exercised by the members of a social stratum on each other. These more or less uniform accumulations of norms are ranked according to their prestige, with the standard language and the elite's dialect

occupying the top of the pyramid. Of course, here as elsewhere, only *mixtures* of meshworks and hierarchies are found in reality, and any given dialect likely belongs simultaneously to a vertical hierarchy and to a horizontal continuum.

The acceleration of city building in the years 1000–1300 affected in many ways the linguistic materials that had accumulated in Europe in the previous millennium. In those three centuries the Romance languages were crystallizing into the forms with which we are today familiar. These stable entities emerged from the continuum of spoken-Latin dialects which coexisted with the standard written form in all the areas that had been subjected to the imperial rule of Rome. In terms of prestige, the homogenized standard was clearly at the top (and would continue to be until the seventeenth century), but social superiority did not translate into linguistic productivity: the written form, precisely because of its much-admired "frozen" body of norms, was largely sterile, incapable of giving birth to new languages. The meshwork of "vulgar" Latin, on the other hand, contained sounds, words, and constructions that replicated *with variation* and were therefore capable of fueling linguistic selection processes and generating new structures. As the sociolinguist Alberto Varvaro puts it, the divergence of the dialects that would become Romance languages began centuries earlier and was kept in check only by the power of the prestigious spoken norm of Rome:

> In Imperial times the linguistic world of Latin had several important properties: a minority endowed with enormous political, social, economic and cultural prestige was absorbing a large majority who were less and less convinced of their own original and diverse identities.... In fact, only Basques and Bretons avoided Latinization; even the Germans, despite the fact that they now held power, gave way to this trend in all the areas where they were not in a majority. Yet, if we go back to the centuries of the Empire, the Latin spoken by these recently Latinized masses undoubtedly tolerated infringement of the norm.... Like all nonstandard phenomena in all languages, some were widely tolerated and some less so, and some were repressed as being too popular (socially and/or geographically).[4]

This state of affairs, in which variation within the meshwork was kept from diverging too much, changed radically with the collapse of the Roman Empire and the concomitant weakening of the hierarchical norm. This resulted, according to Varvaro, in "the loss of the centripetal orientation of the variation."[5] In the centuries leading to the second millennium, only among the feudal and ecclesiastical elites in the different regions was there any sense of "universalism" with respect to the Latin language. The rural masses were left free to reinvent their languages and to forge local identities. The question now is, At what point in time did the speakers of these diverging dialects begin to "feel" they were using different languages? Before the year 1000, with one exception, hardly any of these low-prestige dialects had a definite name or identity. "These forms may have been named by the name of a village or district, when need arose, but more probably never received a name at all."[6] Most likely, all these people perceived themselves as speaking the same language, the spoken version of standard written Latin. Linguistic self-awareness (as well as the names of the new entities) required cultural distance from the linguistic meshwork in which these Latinized masses were immersed and viewing the whole from a hierarchical point of view. Not until the year 813 was the first name for a vulgar variant introduced: "Rustica Romana," which later became vernacular Old French.

This introduction, and the awareness of linguistic divergence that it implied, came in the context of the linguistic reforms that the court of Charlemagne introduced in the ninth century. The specific aim of the Carolingian reforms was to reverse the "erosion" of written Latin, as well as to set standards of pronunciation for the reading of Latin aloud, particularly when reading from the Bible. Unlike the spontaneous evolution of dialects, this act of standardization involved a deliberate act of planning as well as a significant investment of resources (educational, political) to give weight to the new standards:

> The tradition of reading Latin aloud as an artificial language, a sound for each written letter . . . has the air of being obvious, and as though it had been forever present. But someone, somewhere, had to establish that as a standardized norm, for it could not arise naturally in a native Romance community. There was a continuity through the years between Carolingian and Imperial Latin in the vocabulary and syntax of the educated, for these could always be resurrected from classical books by antiquarians, but what we now think of as traditional Latin pronunciation had no such direct continuity with that of the Empire.[7]

The Carolingian reforms were insufficient in themselves to create stable entities with stable names out of the changing "soup" of the dialect continuum, and several other planned interventions were necessary to precipitate the evolution of Romance vernaculars. In the centuries after the reforms, hierarchies of towns began to form with increasing intensity from the eleventh century on, and the local dialects of each of these urban settlements acquired a degree of prestige commensurate to its rank. The most prestigious dialects were those of regional capitals (Florence, Paris) and core gateways (Venice). Simultaneously, the intensification of commercial and governmental activity within these and other towns began to create (or reactivate) a multiplicity of new *uses for written language.* Licenses, certificates, petitions, denunciations, wills, and post-mortem inventories began to be written with increasing frequency and keeping records became part of the daily routine of every merchant or bureaucrat.[8]

At the time of the Carolingian reforms, all four domains of practical literacy—business, government, church, and home—were dominated by standard Latin. But the rise in demand for writing skills forced urban elites, particularly those who spoke the most prestigious dialects, to devise fixed orthographies for their spoken languages and to enforce them as a standard. According to the linguistic historian Richard Wright, writing systems (such as that of Old French) did not evolve spontaneously but were the result of a planned response to specific problems of communication.[9] The development of written forms of the various vernaculars, in turn, acted as a conservative pressure on urban dialects, reducing variation and hence slowing down their evolution. This deceleration may have been perceived by contemporary speakers of a given dialect as the emergence of a stable entity, an impression reinforced by the more or less simultaneous appearance of a name for the written form. But it is not the case that speakers of a dialect had become aware of its divergence from spoken Latin and this awareness provoked them to devise a label for the new language. The divergence did indeed exist as an objective phenomenon, but it was too slow and fuzzy (i.e., Latin diverged into a continuum of dialects) to be directly perceived without an institutional intervention.

The process through which the emerging Romance languages acquired names raises some interesting questions regarding the nature of "naming" in general. According to Gottlob Frege's still-influential theory, the connection between a given name and its referent in the real world is effected through a mental entity (or psychological state) that we call "the meaning" of the name. (Frege called it the "sense" of a name, and Ferdinand de Saussure, his contemporary, called it the "signified.") This meaning, once grasped by a speaker, is supposed to give him or her "instructions" (necessary and sufficient conditions) to identify the object or event that the name refers to. So, for example, the meaning of the words "tiger" or "zebra" allows their users to grasp that which all tigers or zebras have in common (i.e., that which makes them members of that category) and hence endows speakers with the ability to use the names correctly (i.e., to apply them to the right category of entities).[10] The problem here is, of course, that tigers or zebras do not have an essence in common. They are historical constructions, mere agglomerations of adaptive traits that happen to have come together through evolution and acquired stability (at least, enough for us to name them) through reproductive isolation. However genetically homogenized they may be, the external appearance of these animals still reveals a wide range of variation, and, hence, like dialects, they form a continuum of overlapping forms.

A rival theory of reference has been put forth by several philosophers, including Saul Kripke and Hillary Putnam, who deemphasize the "inside the head" aspects of reference and stress the historical and social aspects of language. The basic idea is that all names work like physical labels: they do not refer to an object via a mental entity, but directly, the way the word "this" does. (This is technically expressed by saying that all names have an "indexical component" and hence that they are all like proper names.) Names manage to "stick" to their referents because of the pressures that speakers place on one another: there is a causal chain leading from my use of a word, to the use by the person who taught it to me, to the use by his or her teacher, and so on, all the way to the original "baptismal ceremony" that introduced the label.[11] Hence, one's current usage of a term is "correct" only to the extent that it is connected to the whole *history of uses* of a name. According to this theory, names do not give every speaker the means to specify referents: for many words, only certain experts can confirm the accuracy of the usage. For example, if through genetic engineering we could build animals that looked like tigers or zebras but were a genetically distinct species, the meaning of "tiger" and "zebra" would be of little help to establish correct reference. We would have to rely, as Putnam says, on a social division of linguistic labor which gives groups of experts (geneticists, in this case) the authority to confirm whether or not something is the actual referent of a word, as determined at its baptismal introduction.

Putnam does not deny that we carry certain information in our heads regarding a referent, such as a few identifying traits for tigers (being quadrupedal and carnivorous, being yellow with black stripes, and so on). But these items are in many cases oversimplifications (he calls them "stereotypes"), and far from representing some essence that we grasp, these stereotypes are merely information that we are under a *social obligation* to learn when we acquire the word.[12] Hence, several social factors come into play in explaining how labels "stick" to their referents: the history of the accumulated uses of a word, the role of experts in determining its reference, and the obligatory acquisition of certain information which counts as part of our ability to use the word.

The causal theory of reference may be used to increase our understanding of linguistic history in two different ways. On the one hand, by emphasizing the social practices involved in fixing the reference of a term, nondiscursive practices that intervene in reality become especially important. Thus, successful reference is not purely linguistic and entails expertise in the manipulation and transformation of the objects and events which serve as the referents of words—regardless of whether this expertise is concentrated in a small number of people due to division of labor. In the particular case of the names of the Romance languages, this intervention in reality took the form of expert grammarians assessing degrees of divergence among dialects and devising spelling standards. It also involved institutional enforcement of these standards, resulting in the artificial isolation of some dialects and the consequent increase in their stability and durability. On the other hand, by showing that the meaning of a word is not what allows its users to determine its correct reference, implies that nothing in the meanings of terms like "French dialect" or "French language" (referring to the descendants of Occitan and Francien, respectively) can help us establish some essential difference between them. Our use of the term "French language" would be correct to the extent that it conforms to the history of its uses, a history which began with an institutional baptism, and does not depend on our grasp of some essential features of Francien. (Francien did possess certain distinguishing features, but these features were shared by many nearby dialects and, hence, did not define the essential identity of the dialect of Paris.) In this sense, we may regard the distinction between "dialect" and "language" as completely artificial, drawn by social consensus, and whatever features users associate with the label "French language" (an essential "clarity" or "rationality," for example), as nothing more than a stereotype transmitted through social obligation.[13]

The concept of social obligation is crucial to an understanding of not only naming but language itself. If sounds, words, and constructions are indeed replicators, and if, unlike memes, they do not replicate through imitation but through enforced repetition, then the key question becomes, How exactly are linguistic norms enforced? In what sense are they socially obligatory? The special case of standardized norms offers no difficulty since the enforcement is performed by institutions, including schools and courts and governmental offices, where the standard is used to carry out everyday activities. But what about the population of norms that form the dialect continuum? Sociolinguists answer that, with respect to dialects, it is informal social networks that operate as *enforcement mechanisms.*[14]

To study the social network of a town where a particular dialect is spoken, one would compile for every inhabitant the list of his or her friends, as well as friends of friends. Certain properties of these two circles would then be analyzed: How well do the friends of an individual (and the friends of his or her friends) know one another? Do they interact with each other in multiple capacities (as neighbors, co-workers, kin) or only in specialized circumstances? How likely is it that they will remain within the network after they move up or down the socioeconomic hierarchy? Those networks where there is little social mobility and where the members depend on each other socially or economically are called "high-density" (or "closed") networks.[15]

Small medieval towns and villages would likely have been populated by one or more high-density networks, and closed networks still exist in working-class and ethnic communities in modern cities. On the other hand, those towns in the Middle Ages where a middle class was forming and social mobility increasing were characterized by low-density (or "open") networks. (Needless to say, any given town may contain both extremes and a variety of networks of intermediate density.) For our purposes here, what matters is that high-density networks act as efficient mechanisms for enforcing social obligations. An individual belonging to such a communication net depends on other members not only for symbolic exchanges but also for the exchange of goods and services. The only way to preserve one's position in a network, and hence to enjoy these rights, is to honor one's obligations, and the fact that everyone knows each other means that any violation of a group norm quickly becomes common knowledge. In short, density itself allows a network to impose normative consensus on its members.

High-density networks are especially important to sociolinguistics because they provide researchers with answers to the question of how local dialects are able to survive despite the pressures of an institutional standard. (How, for example, have so many dialects of French survived to this day when the mass media and the system of compulsory education relentlessly promote standard French?) The answer is that language conveys not only referential information but information about group-membership. The sounds, lexicon, and grammatical patterns characteristic of a local dialect are part of the shared values that bind the members of a dense network together and hence communicate information about solidarity and loyalty. In technical terms, the replicators that characterize the dialect of a dense network are said to be transmitted as a highly *focused* set of norms, while the dialects of the upwardly mobile middle classes flow as more *diffuse* sets of norms. Paradoxically, the groups in the very top social stratum (where, by definition, no upward mobility is possible) form dense networks, too, and the norms of their dialects are also highly focused. The difference is, of course, that the norms of elite dialects are highly prestigious while those of local dialects are not, and may even be socially stigmatized.[16] The other difference is that elites, after making their dialects the standards, have access to the institutional means to impose their norms on a much wider speech community, particularly on those with aspirations of upward mobility whose diffuse linguistic norms are prone to succumb to standardization.

The notion of an informal social network is also helpful in understanding the role that individuals (and the stylistic variations to which these individuals give rise) play in the evolution of language. As Labov notes, a given individual variant does not enter this evolutionary process until it has stabilized in a portion of a communication network—that is, until it has become *collective*. In other words, the source of linguistic change is not the idiosyncratic habits of an individual (and certainly not what goes on inside his or her head) but a variant pattern shared by a group and used to communicate with other groups.[17] From this point of view, speakers are not evaluated according to their individual psychological properties but according to the properties of the linkages that bind them to one another.[18] Given a network of a certain density, the higher the local prestige of an individual, or the larger the number of his or her contacts, the more likely it is that a variant originated by that individual will become collective and eventually become part of the accumulated heritage.

In summary, we may picture medieval Europe as a large population of replicating linguistic norms undergoing a variety of transformations and selection pressures: becoming more focused in some areas and more diffuse in others, retaining a meshwork of connections in some parts while elsewhere breaking down into hierarchies around prominent urban centers. Some of these accumulations became consolidated through isolation, becoming more internally homogeneous, while others retained a higher degree of heterogeneity by coexisting with other dialects in different types of *contact situations*. The study of contact between languages is important in historical linguistics because it brings to light the different forms of horizontal flow between dialects, as opposed to the vertical flow of norms through generations. In addition to the flow of linguistic materials between neighboring dialects in a continuum, language may be affected by flows of nonlinguistic materials, such as the migration of a population of speakers who are the organic substratum of a dialect. As we saw before, current maps of the geographical distribution of languages coincide in many parts with genetic maps — not because genes determine languages, but because both travel together during migrations, as well as during colonization and conquest.

The different contact situations created by migratory movements are exemplified by the birth of the English language in the centuries leading to the second millennium. The basic linguistic materials out of which English evolved were first brought to Britain in the fifth century by Teutonic invaders (Jutes, Angles, Saxons) who displaced its original inhabitants, the Celts. Although the Celts were not exterminated (only driven westward) they were largely replaced in most areas of the island without much intermixture. In most cases, the direction of linguistic flow is from the conqueror to the conquered's language; consequently the flow of Celtic norms into the language of the invaders was minimal. In the following six centuries, on the other hand, the basic raw materials provided by the Anglo-Saxon dialects came into contact with several other languages (Latin, several Scandinavian dialects, Norman French), which influenced their evolution in a more dramatic way. Some Latin terms flowed into England from continental Europe as part of the military, economic, and social traffic between Romans and Teutons. But the real influence of Latin norms on the "soup" of Germanic replicators came at the end of the sixth century, when Pope Gregory the Great commissioned Saint Augustine "to lead a missionary band of forty monks in a peaceful invasion of Britain for the purpose of turning the warlike Teutons away from their pagan customs, heathen beliefs, and vengeful practices."[19] The Christianization of Britain (or rather, a re-Christianization, since there were already native Celtic Christians) caused not only a large flow of Latin words to Old English, but also promoted the creation of schools and a system of writing.[20] Conversion to Christianity was effected here, as on the Continent, not by converting each individual inhabitant but by the more efficient procedure of first bringing the ruling elites into the fold. Hence, the flow of words from Latin penetrated the language from the top and flowed downward. The next great influx of alien norms into the still mostly Germanic meshwork of dialects, took the opposite route, penetrating Old English from the bottom up. This was due to several waves of Scandinavian invasions that took place from the eighth to the eleventh centuries. Although as turbulent militarily as those staged earlier by Teutonic tribes, in the end these invasions resulted in coexistence and intermarriage. In these centuries, Scandinavian words such as "they," "though," and about eight hundred others were added to the mixture.[21]

By the turn of the millennium, Old English had evolved through several types of contact: one caused the replacement of Celtic norms, another fostered coexistence between different Germanic norms, and, in between, still another facilitated a cultural penetration by Latin norms. The transformation of Old English (which is closer to German) into Early Middle English (which is recognizable as "English") took place in yet another contact situation: the wholesale replacement of the local elite by a foreign one. In the eleventh century, as the different dialects of French were finalizing their differentiation from Latin, the French-speaking Normans staged a successful invasion of England and ruled that country for nearly a century (1066–1154). The Old English-speaking nobility virtually ceased to exist, and even the highest offices of the church fell into Norman hands. French became the language of the elites for over two centuries, while Old English became the low-prestige dialect of the peasant masses. In this way, the Norman Conquest affected Old English much the way the collapse of the Roman Empire affected Latin, as we observed earlier. As one historian puts it:

> The most important single influence of the Norman Conquest upon English was the removal of the conservative pressures that tended to impede its evolution. As the tongue of a subjugated country Old English lost prestige. West Saxon was no longer the literary standard of the conquered Britons, and the Anglo-Saxon scribal tradition was suppressed. Neither church nor state had much time to give to the language of the English peasants, and the socially and intellectually elite could not be bothered with it. Under such conditions of laissez faire, the language benefited from a return to oral primacy: colloquial use determined usage and variant dialect forms competed for acceptance. Unhindered by rules of prescription and proscription, the English peasants . . . remodeled the language with tongue and palate.[22]

Thus, thanks to the forceful removal of an emerging standard (West Saxon), the flow of norms through several generations of English peasants became more fluid, the amount of variation increased, and the whole continuum of dialects evolved faster. By the time the English elites rediscovered their native language in the thirteenth century, it had already changed in dramatic ways. In particular, it had been transformed from a *synthetic* language into a mostly *analytic* one. These terms refer to alternative ways in which languages express certain grammatical functions. A synthetic language expresses functions like the gender and number of nouns, or the tense of verbs, via certain linguistic particles called inflections. Modern English retains a few of these (the *-s* for plural and the *-ed* for past tense), but most of the inflections from Old English have been dropped. Inflectioned languages are free to position words in sentences in several alternative ways (since they carry grammatical markers with them), while languages that have lost their inflections express grammatical functions through a fixed word order (e.g., subject-verb-object). Given that word order captures very economically the logic behind a sentence, these languages are called analytic.

Ethnocentric linguists in the past (particularly those studying English and French) didn't see in the transformation from synthetic to analytic a simple switch from one set of grammatical resources to another *equivalent* one, but rather a move up the ladder of progress, as if an internal drive for greater clarity (rationality) were guiding the evolution of languages. But similar grammatical simplifications occur in languages that chauvinistic speakers of English or French would never consider to be on the same level as their mother tongue. These are the so-called *trade jargons*, or *pidgins*, like the famous Sabir, or Mediterranean lingua franca, a long-lived dialect widely used in the Levant trade beginning in the Middle Ages. The study of pidgins is particularly relevant here not only for the light it throws on the distinction between analytic and synthetic, but also because it illustrates yet another type of contact situation that affects linguistic evolution: the transitory linguistic contact created by military or trade encounters between alien cultures.

The origins of Sabir are obscure. One theory postulates that it was born of the Crusades, beginning in the year 1095. If so, the Jerusalem battlefields would have been its place of birth, from whence it spread following military and merchant movements.[23] Critics of this theory point out that as late as the thirteenth century many Levant trade documents were written not in Sabir but in a changing hybrid of Italian, French, and Latin. Sabir may have emerged shortly after, and then, thanks to its simplicity, replaced those early hybrids. On the other hand, it may never have existed as a single entity but as a series of pidgins, each drawing on different Romance languages for its lexical materials.[24] For example, in the early Middle Ages the vocabulary of Sabir may have relied mostly on borrowings from the dialects of Genoa and Venice, since those cities dominated trade with the Levant. When later on the Portuguese found alternate routes to the luxury markets and began to break the monopoly of the Italian cities, Sabir's vocabulary changed accordingly. At any event, Sabir is rare among pidgins because of its longevity (it died only in the early twentieth century, as the Ottoman Empire collapsed). Most pidgins emerge and disappear as the short-lived contact situations that give rise to them come to an end. But pidgins endure wherever contact between alien cultures has been institutionalized, as happened, for example, at slave trading posts and on sugar plantations.

One distinctive feature of pidgins — what differentiates them from simple mixtures — is that they greatly simplify the set of norms from which they were derived. Many redundant features of languages (such as the verb "to be") are eliminated, since their main function is to make speech more self-contained or redundant (i.e., less dependent on contextual clues for correct interpretation). Without these resources, pidgins become more dependent on context, so that, in a sense, behavioral acts such as pointing to referents become part of the "grammar" of the pidgin. Yet, far from being degenerate tongues that devolved from their "master" languages, pidgins are creative adaptations of linguistic resources.[25] Slave pidgins, for example, were not a kind of "baby talk" created by the master to communicate with his slaves, but a creative adaptation by slaves from disparate linguistic backgrounds to communicate with one another.[26]

Due to their stigmatization as "inferior" languages, pidgins did not become a serious subject of study until relatively recently. Today, the field is growing explosively as ethnocentric prejudice gives way to a more objective approach. Simultaneously, the emphasis has changed, and linguists are less interested in pidgins as distinct entities than in "pidginization" as a general process that may or may not create a stable entity. Before this switch in approach, the creation of stable entities was seen as a simple process consisting of two successive stages: first, a "target" language (e.g., the language of the slave master) was simplified and a pidgin was created. Then, when the slaves were set free, the first generation of children who learned the pidgin as a mother tongue re-created many of the redundant features that had been stripped away, and a new entity emerged: a *creole*. (Of course, not only children participate in this recomplexification of the pidgin; adult speakers may also contribute by borrowing items from other dialects.)[27] Although this process of crystallization of new creole languages via enrichment of a pidgin is still of great interest to linguists (since it represents an accelerated version of linguistic evolution, one that is compressed into one or two generations), today's emphasis is more on the processes of pidginization and creolization in general, whether they result in new stable entities or not:

> A linear model of two discrete steps, as implied by the standard conception of pidgin and creole, may oversimplify the complexity of the historical cases to the point of distortion, and in itself contribute to the difficulty of inter-

preting the evidence. Within a single region there may coexist, contiguously, more than one stage of development. And there may indeed be more than two stages—a pre-pidgin continuum, a crystalized pidgin, a pidgin undergoing de-pidginization (reabsorption by its dominant source), a pidgin undergoing creolization, a creole, a creole undergoing de-creolization.[28]

A number of linguists and philosophers of language have noted the similarity between the contact situations giving rise to these processes and those behind the emergence of the Romance languages and English. This is not to say that the Romance languages or English should be considered pidgins or creoles, but they may also have undergone simplifications and recomplexifications. For instance, the loss of inflection and the fixing of word order which distinguish analytic languages such as French and English can also be observed in the evolution of many pidgins. The removal of a dominant norm (West Saxon in the case of Old English, Roman Latin in the case of Old French), which increases variation and hence the speed of divergent evolution, is also a constant factor in the development of pidginized languages. On the other hand, the expanding vocabulary and multiplying uses of language (in education, law, etc.) that characterize creoles are also part of the birth process of dominant languages (as when Parisian French replaced Latin or when London's English replaced Norman French).[29] Thus, the population of linguistic replicators that inhabited Europe in the Middle Ages may be seen as having undergone processes not only of focusing and diffusion (in social networks) and hierarchization (in urban centers) but also of pidginization and creolization.

Such is, in so many words, the linguistic viewpoint adopted by Gilles Deleuze and Félix Guattari, who call those languages that have risen to the top of a hierarchy "major" languages and those forming a meshwork of dialects "minor" languages. Yet they do not use these terms to refer primarily to stable entities (some more homogeneous, some more heterogeneous) but rather to the processes (becoming major, becoming minor) that affect the population of norms as a whole:

> Should we identify major and minor languages on the basis of regional situations of bilingualism or multilingualism including at least one dominant language and one dominated language...? At least two things prevent us from adopting this point of view.... When [modern] French lost its worldwide major function it lost nothing of its constancy and homogeneity. Conversely, Afrikaans attained homogeneity when it was a locally minor language struggling against [modern] English.... It is difficult to see how the upholders of a minor language can operate if not by giving it (if only by

writing in it) a constancy and homogeneity making it a locally major language capable of forcing official recognition.... But the opposite argument seems more compelling: the more a language has or acquires the characteristics of a major language, the more it is affected by continuous variations that transpose it into a "minor" language.... For if a language such as British English or American English is major on a world scale, it is necessarily worked upon by all the minorities of the world, using very diverse procedures of variation. Take the way Gaelic and Irish English set English in variation. Or the way Black English and any number of "ghetto languages" set American English in variation, to the point that New York is virtually a city without a language.[30]

To return to the Middle Ages, the accelerated urbanization that produced regional hierarchies of towns created several high-prestige vernaculars for each portion of the continuum of Latinate dialects. Each regional capital witnessed the rise of its own variant to the status of a locally "major" language, which had its own writing system and accumulated prestige at the expense of a number of "minor" variants spoken in low-rank small towns and rural supply areas. Thus, the continuum of French dialects was divided into two regions struggling for supremacy: a family of southern dialects called langue d'oc and another family spoken in the north and center, known as langue d'oïl, which included the Parisian vernacular (Francien) as well as the variant that the Normans had imposed on Britain. Nothing intrinsically linguistic was to determine the outcome of this struggle between langue d'oc and langue d'oïl. On the contrary, the ascendant prestige of langue d'oïl was the result of a variety of nonlinguistic events. The successful colonization of the British Isles by the Normans was one such event, as was the Albigensian Crusade, which benefited Francien at the expense of Occitan, a member of the langue d'oc family. A rather precocious political centralization around Paris added to the momentum, as did extensions in the usage of vernacular, such as the translation of the Bible (into Francien) in the year 1250 by scholars at the University of Paris.[31]

Other emerging Romance languages followed similar lines. On the Iberian Peninsula, several regional variants developed, and Catalan began to diverge from the rest (known collectively as the Hispano-Romance dialects) around the ninth century. The dialect that would eventually rise to the top, Castilian, was at first a rather peripheral variant spoken in the region that later (around 1035) became the Kingdom of Castile. Castilian's potential rivals, Leonese and Aragonese, were at that time more prestigious and more in keeping with the Romance languages spoken outside the peninsula. The rise of Castilian began with the war against Islam, which had colonized the southern regions of the peninsula for eight centuries. The Kingdom of Castile played the most important role in the war of reconquest, beginning with the capture of Toledo in 1085. Through the prestige won during the war, as well as the migration of Castilians to settle the reconquered territories, the cultural and territorial influence of Castilian grew at the expense of other Hispano-Romance dialects, most of which, forced to the defensive, eventually withered away.[32] After the reconquest, Toledo's new Castilian-speaking elites, together with those from Seville, furnished the materials from which the Spanish language eventually evolved.

Unlike France and Spain, where political centralization came relatively early, Italy and Germany would remain fragmented for centuries because of the opposition to central rule by their independent city-states. This fragmentation, or rather resistance to homogenization, acted as a linguistic centripetal force. Certain urban vernaculars rose to prominence, but their triumph was less clear-cut and linguistic dominance often shifted between regions. For instance, the dialect of the city of Lübeck became the standard of the powerful Hanseatic League; but when the commercial success of the league waned, other German variants became dominant.[33] In Italy, the Tuscan dialect had enjoyed a privileged status since the fourteenth century; it had been adopted not only by the papal court but by a number of literary writers, which greatly increased its prestige. However, each Italian city-state retained its own local variant for centuries (that is, the variant used by its elites), and linguistic unification was not attempted until the nineteenth century.[34]

Besides these local movements in which a few variants were "becoming major" relative to the rest of the continuum, there was a global struggle between the local major languages and the undisputed global major: written Latin. This struggle, which took place between the thirteenth and eighteenth centuries, is known as the "rise of the vernaculars." Latin, which in the early years of the Roman Empire had been a minor language in comparison to Greek, began the new millennium greatly strengthened, for several reasons. Its role as the official language of the church had been codified in the year 526 with the Benedictine Rule, which gave it a central place in monastic literacy and manuscript production, a status reinforced by the Carolingian reforms. The centralization of religious power and consolidation of ecclesiastical hierarchies between the years 1049 and 1216 allowed the institutionalization of Latin as the obligatory medium for the conduct of mass, while the vernaculars were forbidden from playing this role.[35] Finally, the linguistic heterogeneity prevailing in

Europe created the need for a lingua franca for international communication, and Latin easily eclipsed Sabir and the other low-status pidgins (such as Mozarabic) that may have performed this role.

But the agricultural and commercial intensifications that began complexifying urban life from the eleventh century on soon altered Latin's status. The uses for writing greatly diversified, and the demand for literate individuals greatly increased in administration, law, and commerce. The establishment of cathedral schools and urban universities shifted the center of education toward the new towns and away from rural monasteries. (In Italy there were even some lay schools where the instruction was conducted in the vernacular.) Lay officials gained increasing importance at the expense of the clergy, at least within the world of secular administration. Finally, there were processes affecting not the institutional but the organic substratum of Latin, such as the Black Plague of the fourteenth century. As William McNeill suggests, "The rise of vernacular tongues as a medium for serious writing and the decay of Latin as a *lingua franca* among the educated men of Western Europe was hastened by the die-off of clerics and teachers who knew enough Latin to keep that ancient tongue alive."[36]

The battle between the dominant urban vernaculars and Latin was not a struggle to dominate the tongues of the masses, but rather a struggle to dominate the language of public institutions. The dialects of the lower strata of medieval society were tightly bound up with their speakers and migrated with them and their genes. A dialect's highly focused set of norms is more easily killed (by replacing one population of speakers with another) than absorbed by alien languages. For this reason, while prestige determines the relative position of a dialect in a hierarchy, and hence its short-term destiny, *the sheer weight of numbers* decides its ultimate fate. Norman French, for example, however prestigious it may have been as the official language of the English aristocracy, never had a chance to take over as the language of the English masses.[37]

Similarly, written Latin was in no position to compete with the vernaculars. During the period of rapid urbanization that began in the eleventh century, the population of Europe doubled, and with it the number of vernacular speakers. But the French of the Parisian elites, for example, was never in competition with Latin as a popular language for France but as the official language in French courts, government offices, and places of higher learning. Francien, too, began competing with Latin as the language of international diplomacy. In this case, raw numbers counted less than accumulated prestige: "French's long period of predominance as the major international language of culture and diplomacy long antedates its general use as spoken language in France: by the end of the seven-teenth century, French had in effect replaced Latin in the former role . . . at a time when Francien was the native tongue of perhaps a quarter of the population of France."[38]

Francien had achieved the status of a "norm to aim for" by the thir-teenth century, in terms of unofficial writing and cultivated speech, but it did not overtake Latin until a series of fifteenth- and sixteenth-century edicts, such as the Edict of Villers-Cotterêts of 1539, made its use obliga-tory in official writing.[39] In England, too, we find that certain institutional interventions changed the status of the English language through a series of official acts, such as the Statute of Pleading enacted by Parliament in 1362, which made English the official language of the British courts. Court records, however, were still kept in Latin, and the statute itself was written in French. Yet, by 1489, "Henry VII put an absolute end to the use of French in the statutes of England. With that act the language that had gone underground in 1066 emerged completely triumphant over for-eign domination."[40] These official acts, which transformed the status of English, French, and Latin more or less "instantaneously," are special cases of what the "ordinary language" philosopher J. L. Austin called "speech acts": social actions performed by the very utterance of a string of words. Commands, such as the order to use English or French in cer-tain official contexts, are one type of speech act. The making of promises or bets, the issuance of warnings, verdicts, or judicial sentences, the baptizing of an object or a person, and many other verbal actions that carry with them social obligations and consequences are also examples of speech acts.

According to Austin, speech acts involve a conventional procedure that has a certain conventional effect, and the procedure itself must be exe-cuted correctly, completely, and by the correct persons under the right circumstances.[41] The declaration of English as the official language of government, for instance, had to be made by a person with the authority to issue such declarations and in the right institutional setting. Not just any utterance of the words "I declare you the official language" carries the illocutionary force of a command. This simply emphasizes the fact that we are not dealing here with a purely linguistic process but with a complex situation involving hierarchies, chains of command, and the means to enforce obedience. Austin distinguishes those speech acts performed in judicial courts (and other institutional settings), where the procedure is so routinized that what counts as "correct" is clear to every-one, from those speech acts used in everyday life, where the procedures are not rigid or formal and where, therefore, there is more room for ambiguity. Nevertheless, as we saw above, communication networks may act as enforcement mechanisms for promises or orders even in the absence of explicit criteria for the correct performance of a speech act.

We may compare the instantaneous transformations in status which a command, guilty verdict, or death sentence effect with the phase transitions that materials undergo at certain critical points. Much as liquid water suddenly switches from one stable state to another and begins to become solid ice when the temperature or pressure reaches a particular threshold, so a guilty verdict may abruptly change the social status of a person, who will be switched from a state of free motion to one of confinement. However fruitful this comparison may be, at the very least it calls attention to the fact that much as genetic replicators impinge on the world as catalysts for chemical phase transitions, so linguistic replicators affect reality by catalyzing certain "social phase transitions."[42]

In addition to the official speech acts that abruptly changed their status, the dominant vernaculars of each region needed to enrich their reservoirs of expressive resources in order to effectively challenge the international standard. No official declaration could have made French or English the official medium in which to conduct government business if their vocabularies had not contained all the technical words required in judicial, legislative, diplomatic, military, and administrative communications. One means of increasing vocabulary was to use these languages' word-forming resources to generate the needed lexical items. Literature played a key role in this respect, enriching the expressive resources of the ascendant dialects while increasing their cultural prestige.

The ascendant dialects also expanded their lexicons by borrowing words from other languages and then adapting the borrowings to local usage. These linguistic flows from one population of norms to another display some interesting patterns that illuminate a number of the internal features of language. For instance, although the individual words of a language are free to replicate from one culture to another as memes (that is, by imitation or borrowing), a language's sounds and grammatical patterns, particularly those that are central to a language's (historical) identity, tend to move together with its speakers. Furthermore, words related to questions of everyday survival, unlike technical or literary words, do not diffuse well among different languages.

Modern English, for instance, still contains an archaic residue of Old English words, surrounded by the vast cosmopolitan vocabulary that it accumulated slowly, via diffusion (i.e., via various flows of memes). The words "father," "mother," "child," "brother," "meat," and "drink," as well as those that express basic activities such as "to eat," "to sleep," "to love," and "to fight," derive directly from the Germanic vocabulary of Old English. On the other hand, most of the technical vocabulary for ecclesiastical matters flowed into English from Latin during the period of Christianization. (About 450 Latin words were introduced into English during this period.) Military, legal, governmental, and medical terms (as well as some culinary and fashion vocabulary) entered the English reservoir in large numbers (about ten thousand French words) during the Norman occupation. Soon after the occupation ended and English military victories made the French seem less of a threat, large quantities of Parisian French words began to flow into Britain, peaking in intensity between the years 1350 and 1400.[43] The direction of this flow of memes ran from the language that had accumulated more prestige and lexical complexity to the less prestigious and complex one. This is, of course, a relative distinction: while French was for a long time more culturally prestigious than English, during the fifteenth and sixteenth centuries it was "inferior" to Spanish and Italian and many Spanish and Italian words flowed into France from those two countries.[44]

The many hundreds of French words that flowed into Middle English suffered different fates. Some of them were simply taken as they were, but many were assimilated into local dialects. Borrowed French and Latin words often coexisted with their English synonyms, instead of displacing one another or hybridizing. In the fifteenth century English developed a trilevel system of synonyms with different levels of prestige: commonplace English ("rise," "ask"), literary French ("mount," "question"), and learned Latin ("ascend," "interrogate"). As one historian puts it, this accumulation of synonyms allowed "for a greater differentiation of styles — in both formal and informal usage.... Thus the native English vocabulary is more emotional and informal, whereas the imported French synonyms are more intellectual and formal. The warmth and force of the former contrasts with the coolness and clarity of the latter. If a speaker can be intimate, blunt, and direct in basic English, he can also be discreet, polite, and courteously elegant in the diction of borrowed French."[45]

This hierarchy of synonyms is a special case of what sociolinguists call "stylistic stratification," that is, the ranking of a language's different *registers*, which are reserved for use in particular situations: a casual register, to be used with friends and family; a formal register, which is used, for example, in institutional situations or simply when talking to strangers or superiors; and a technical register, used at work or when communicating with other professionals. Of course, the vocabularies of these registers need not come from different languages. The distinction is drawn more in terms of the amount of care that one puts into the creation of

sentences during a linguistic exchange (or, in the case of technical registers, by the use of special vocabularies or technical jargon).[46]

English speakers in the Middle Ages and Renaissance presumably engaged in register switching according to the degree of formality of a situation. Outside of London, they likely also engaged in a related process
called code switching. Due to geographic isolation, the flow of linguistic
replicators that made up Old English had generated five different "species"
of Middle English (Southern, Kentish, East Midland, West Midland, and
Northumbrian). While the dialect of London had by the fifteenth century
become the most prestigious form of English, it did not replace the other
dialects but, rather, was added to the population as a *superimposed norm*.
This meant, for instance, that a speaker of Kentish who also knew the
London dialect would indeed switch codes when talking to different people, using a local code in talking to a neighbor and an interregional code
in addressing someone from the capital. Other countries, such as Italy
and Germany, where political unification came late, remained much more
linguistically fragmented; consequently, their inhabitants practiced code
switching on an even more extensive basis.[47]

Code and register switching are further examples of contact between
different dialects, a kind of "internal contact" that tends to make them
less internally homogeneous. Indeed, when one compares any actual language's internal variety—keeping an eye on its coexisting registers and
codes—with "language" as imagined by structural linguists and semioticians, the most striking difference is the high degree of homogeneity
that linguistic theorists take for granted. The semiotician seems to always
have in mind a simple communication between a speaker and a listener,
wherein both speak precisely the same language with identical skill.
This oversimplification becomes all the more obvious when one studies
countries where stable bilingualism is the norm, such as Belgium or
Canada, not to mention India, which today recognizes fourteen official
languages. In the Middle Ages and the Renaissance it was not uncommon for people to be multilingual: Christopher Columbus, for example,
spoke Genoese as his mother tongue, wrote some Latin, and later
learned Portuguese and Spanish.[48] As Labov stresses, command of a real
language, unlike the simplistic characterization of linguistic competence
made by the structuralist school, involves the ability to deal with great
amounts of heterogeneity.

Hence, behind any uniform set of linguistic norms there must be a
definite historical process that created that uniformity. The processes of
homogenization that were at work on the Indo-European dialects that
became the Romance and English languages may be said to have come

in two great waves. The first wave took place as part of the general process of urbanization: the ascendancy of the London and Paris (and
other) dialects to the top of the linguistic hierarchy, leading to their adoption as official languages of government communication and lower education. This first wave involved both unplanned processes (including
positive feedback; for instance, the more literature appeared in a given
dialect, the more viable a literary medium that dialect seemed to other
writers) and institutional speech acts that triggered sharp transitions in
the status of certain vernaculars. Other than the effort to create writing
systems for the elite dialects, the first wave did not involve great
amounts of linguistic "self-awareness," that is, conscious analysis of
the internal resources of a language and deliberate policies to extend or
fix those resources. The sixteenth and seventeenth centuries, however,
witnessed the emergence of the first efforts at what we would today call
"linguistic engineering." The second wave of homogenization involved
institutional policies aimed at the deliberate "slowing down or complete
stoppage of linguistic change," or, in other words, "the fixation forever
of a uniform norm."[49] That this goal has turned out to be unattainable
in practice (to this day minority languages thrive alongside the standards) does not mean that the institutional enterprises that Spain, Italy,
and France embarked upon during this period did not have great historical consequences.

The second wave may be said to have begun in Spain, when for the
first time the grammar of a Romance dialect (Castilian) was systematically set forth. Unlike written Latin, which as a "dead" language had to
be transmitted in schools by means of *explicit rules*, the various regional
dialects of Spain were learned at home as one's mother tongue. The
grammarians of the Renaissance did not discover the "real" rules of language (not even Chomskyans today have achieved this), and they did not
claim to have done so. Elio Antonio de Nebrija, who published the first
grammar of Castilian fifteen days after Columbus had sailed to "discover"
America, was quite conscious that his invention was an artifact ("artificial
Castilian" he called it[50]), but one that had great potential as an instrument of homogenization. As the sociolinguist Elnar Haugen writes, "The
close connection of grammar and politics is shown in the fact that the
first Spanish grammar appeared in 1492 and was dedicated to Queen
Isabella; it was intended to be a companion of the Empire, the author wrote,
and should spread Spanish [i.e., the Castilian dialect] along with the rule
of the Spaniards."[51]

According to Ivan Illich, both Columbus and Nebrija came to the queen
to propose complementary projects: one to extend royal power into new

lands, the other to increase the inner cohesiveness of the sovereign body via a homogeneous language. Unlike classical Latin, which had been "engineered" so that the speech patterns of Roman senators and scribes could be regulated, the target of Nebrija's proposed reforms was not the language of the Spanish elites but the unbound and ungoverned language of the masses. Moreover, to the extent that the multiplicity of dialects learned informally at home were superseded by an artificial ("Castilian") language taught formally, like Latin, as a set of rules, Nebrija's grammar was the first step toward what centuries later would become a compulsory education system based on a standardized language. In a way, as Illich remarks, this meant replacing the autonomous linguistic resources of dialect speakers with a reservoir controlled by institutions and given to the masses as a gift from above.[52] In the end, Nebrija's project failed to gain institutional support from the royal court, but the same concern with creating artificial languages that would be "pure" and "long lasting" would reappear elsewhere in different forms.

In Italy, for example, the Tuscan (i.e., Florentine) dialect had come to play the same dominant role as the Castilian, Parisian, and London dialects. Tuscan had been "creolized" (enriched) by several writers (Dante, Boccaccio, Petrarch) who not only enlarged its reservoir of expressive resources but also increased its prestige relative to the dialects of other important cities (Venice, Genoa, Milan). In 1582, the first institution specifically designed to act as a brake on linguistic change was born in Florence: the Academy of Language, an organization dedicated to the creation and dissemination of artificial Tuscan through the publication of grammars, dictionaries, orthographies, and other formal codifications of language.[53] This project, like Nebrija's, proved hard to achieve in practice, particularly because the political strength of the city-states retarded national unification until the nineteenth century.

Still, the Florentine Academy of Language had a more concrete influence, inspiring the creation of similar institutions in nascent nation-states such as France, where an organization modeled on the Italian paradigm was born in 1637 as part of Richelieu's plan to unify the country. The French Academy had as its explicit mandate the purification and perpetuation of the French language, or as one of its members put it, to "fix language somehow and render it durable."[54] By 1705 the academy could boast that if only the words included in its official dictionary were used, French would remain fixed for all time.

This second wave of homogenization, like the first one, did not produce master languages that completely replaced the dialect continua of their respective countries. The academies simply added one more set of norms to the existing population, a new set with a hierarchical structure super-imposed on the meshwork of dialects. As the French linguist Antoine Meillet said, standard French "has never been the language of any but a few people and is today not the spoken language of anyone."[55] The new artificial rules of grammar and spelling, the pyramidal vocabularies con-tained in dictionaries, and the other devices of "linguistic engineering" (such as books on rhetoric and poetics) affected most of all the formal register of the languages in question, leaving the casual register mostly untouched. (The technical register of French would not be affected until the eighteenth century, when Lavoisier and others helped fix the way in which suffixes and prefixes should be used to coin new scientific terms.) However, it was precisely the formal register that needed to be standard-ized if the vernaculars were to triumph over Latin. Hence, in the general process of the rise of the vernaculars, standardization did have a lasting impact. The other decisive element in this linguistic war was provided by technology: the printing press.

Although the concept of movable type may not have originated with Johannes Gutenberg (there are Chinese, Korean, and even Dutch ante-cedents), he was certainly the first to implement a practical way of automating writing. Several technical problems were solved during the 1440s (adjustable molds for casting durable type and a special ink suit-able for metal type were developed), which enabled Gutenberg to create a machine that, when meshed with the burgeoning paper industry, brought down the cost of reproducing texts considerably and allowed the true mass dissemination of the written word. Of the twenty-four thousand non-Greek works printed in Europe before 1500, about 77 percent were in Latin, the rest in vernacular. But the number of works printed in the vernaculars increased over the years and the vernaculars predominated by the end of the seventeenth century.[56] The Protestant Reformation, by championing the translation of the Bible into vernaculars, dealt a power-ful blow to Latin's domination of ecclesiastical rituals and, more impor-tantly, education. Thus, in one sense, the printing press aided some minor languages in their struggle against a major language. And yet, given that the major-minor distinction is entirely relative, the printing press simulta-neously aided locally major languages (the rising standards) in their struggles against potential local rivals.

Moreover, since the very existence of a writing system exerts a homog-enizing influence on a language and acts as a brake on linguistic change, the mechanical reproduction of texts amplified in several ways this con-servative trend. In England, where William Caxton introduced the printing press in 1476, the printed word promoted the written standard of the elite London dialect as a brake on linguistic variation. As the historian John Nist has written, "Along with extending literacy and expanding popular education, the printing press became a powerful cultural force that put back into the language what had been lost with the Norman Conquest —the conservative pressures of self-awareness and social snobbery."[57] English printers, on the other hand, locked into type certain spelling rules that did not entirely correspond to the phonemes of English, sounds that were, at any rate, changing as these norms were being frozen. And yet, as Nist puts it:

> More important than either the orthographic conservatism or the phonological inconsistency wrought by the printing press was the mistaken notion that English is primarily the written word. The grapheme and the visual morpheme began to dominate the literary imagination, and the raw power of the oral tradition gradually gave way to the elegant refinement of the silent literary. In time, the divorce between the spoken and the written was legalized by the authoritarian grammarians of the eighteenth century and their heirs.[58]

I would like to conclude this section with a brief description of those processes affecting linguistic evolution which are *internal* to language. For example, at the very same time that printers and grammarians were attempting to freeze set correspondences between sounds and written signs into a spelling standard, the English language was undergoing a dramatic change in its sound system. This transition, which involved several generations of speakers, goes by the name of the Great Vowel Shift:

> When Chaucer died in 1400, people still pronounced the *e* on the end of words. One hundred years later not only had it become silent, but scholars were evidently unaware that it ever *had* been pronounced.... [Thus] in a relatively short period the long vowel sounds of English... changed their values in a fundamental and seemingly systematic way, each of them moving forward and upward in the mouth. There was evidently a chain reaction in which each shifting vowel pushed the next one forward: The "o" sound of *spot* became the "a" sound of *spat*, while *spat* became *speet*, *speet* became *spate*, and so on. The "aw" sound of *law* became the "oh" sound of *close*, which in turn became the "oo" sound of *food*. Chaucer's *lyf*, pronounced "leef," became Shakespeare's *life*, pronounced "lafe," became our "life." Not all vowels were affected. The short *e* of *bed* and the short *i* of *sit*, for instance, were unmoved, so that

we pronounce those words today just as the Venerable Bede said them twelve hundred years ago.[59]

No one is exactly sure what started this "chain reaction" of shifting vowels. It could have been an articulatory shortcut, in which the "least effort" principle favored the stabilization of a new sound in a given speech community; it could have also been a mere mistake in pronunciation which spread by imitation; or, finally, it could have been a new variant sound introduced into a community through one of the many different kinds of contact situation. In a way, the trigger for the Great Vowel Shift is its least important aspect compared with the dynamical changes unleashed by the catalyst. Given that there is no intrinsic connection between the sounds that make up a word and the meaning (or obligatory semantic information) carried by the word, the usefulness of a given set of sounds is guaranteed by the more or less systematic contrasts that they have with one another. If one of the sounds moves toward another, thereby reducing the contrastive power of both, the second sound must move as well. This "push-chain" dynamic then continues until a whole series of sounds has acquired a new position that preserves the original contrasts. Simultaneously, the "empty space" left behind by the very first movement may now trigger another series of motions by an unrelated series of sounds to "fill" that empty slot. Linguists call this secondary reaction "drag chain" dynamics.[60]

The fact that these internal rearrangements occurred largely unconsciously over several generations could mislead us into thinking that they were the product of an internal drive in language. Although completely circular shifts like this one may be considered "homeostatic mechanisms" (and may be said to endow the system of sounds with a certain degree of autonomy from grammar, vocabulary, and social pressures), they can be explained using the same mechanism that explains other (less autonomous) changes in language: an interplay of variable linguistic replicators and the sorting device constituted by selection pressures (in this case, the need to preserve the functionality of language in everyday communication tasks).[61] Moreover, push- and drag-chain dynamics and, more generally, slow switches from one stable state to another may occur not only in the sonic substance of the spoken chain, but also in the realms of vocabulary and syntax.

For example, certain words (such as the verbs "to get" or "to do") may become slowly emptied of their lexical meaning and become "grammaticalized," that is, selected to become relatively "meaningless" particles used to express grammatical functions. The desemantization of words as a means of recruiting new grammatical devices is a slow and unconscious process and provides us with yet one more source of heterogeneity.

This is, in fact, the type of heterogeneity that Labov stresses the most: the existence in a language of *variable rules*.[62] A good example is provided by the grammaticalization of the verb "to do," which was recruited as a device to express negative and interrogative clauses. Its desemantization occurred slowly, beginning in the thirteenth century, but it remained only a peripheral grammatical device until the end of the fifteenth. Then, during the years 1535–1625 it was pressed into service to perform an increasing number of syntactical functions, later on decreasing in range until settling into the role it plays today. The important point here is that, despite its growing range of functions, "it was by no means obligatory in them at the end of the sixteenth century (e.g. *goest thou, he goeth not* were still common), while in affirmative clauses it was ... in free variation with the simple verb forms for the expression of tense."[63]

Today, of course, the use of "to do" is obligatory to express some grammatical functions in English, which means that over a period of several centuries the grammatical rules for the use of this desemantized particle transmuted from optional and variable to categorical. According to Labov, linguistic competence should be defined in such a way as to include the ability to handle these variable rules, at different states of their evolution. Moreover, he attacks the tradition (among Saussureans and Chomskyans) of concentrating on a study of standard languages precisely because their artificial homogeneity obscures the existence of nonuniform, changing grammatical devices. (Labov, for instance, finds a variety of variable rules in his study of Black English—rules that do not exist at all in standard American English.[64]) When we add this internal, systematic source of variation to all the other sources that we have examined so far, the picture of language that emerges is one of a heterogeneous mixture of norms in constant change, very different from the traditional view of a timeless, universal structure isolated in its "synchronic" heaven from all the turmoil around it. As Deleuze and Guattari put it: "You will never find a homogenous system that is not already affected by a regulated, continuous, immanent process of variation (why does Chomsky pretend not to understand this?)."[65]

Furthermore, this variable soup of linguistic (replicating and catalyzing) materials was constantly intermingling with all the other material and energetic flows that we have examined in this book. Cities, particularly large cities, were the places where the strangest mixtures of food and genes, money and words, were concocted. The intensity of trade, which contributed to social mobility (and the creation of a middle class), detached some people from their original communication networks (and from dependence on relatives and neighbors for their livelihood), decreasing the conservative pressures that group loyalty put on linguistic change, and allowing the downward penetration of the standard. Also, middleclass speakers, in their anxious usage of the high-prestige variant in their now more impersonal and fragmented social networks, tended to "hypercorrect" their dialectal speech, adding an additional source of variation and heterogeneity.[66] On the other hand, the constant flow of rural immigrants which kept cities alive and growing also brought in linguistic materials that contributed much to the formation of ghetto dialects.[67] Large cities, therefore, contributed not only to a defocusing of the norms (by prying open social networks via upward mobility) but also to the creation of new closed networks and, hence, new focused ethnic variants:

> Large cities bring together the critical mass of similar people needed to found communities. While the Irish in small Leicestershire villages were forced to blend in with the native English, those in Glasgow began Catholic churches and clubs, building communities around their ethnic loyalties.... Large cities ... produce strongly articulated value systems rather than isolated individuals. They are not melting pots, but mosaics of disparate groups, each of which fights to maintain its own identity. At first glance, this view of cities is puzzling, for how can a place be both impersonal and culturally intense? How can an individual be both anonymous and closely involved in a specific subculture? The answer is that cities contain both large-scale and small-scale environments. Although in public places — the stores, offices, streets, and large institutions — contacts are relatively brief and anonymous, there is a separate, private social life to be found on the level of family, neighborhood, club, and ethnic group that operates with different rules.[68]

Urban centers, by housing dynamical mixtures of energy, matter, and catalytic replicators of different kinds (genes, memes, norms, routines), greatly influenced linguistic evolution before the seventeenth century. After that they would continue to play important roles, but now as part of larger sociopolitical entities: as the capitals of the emerging nationstates. While before the French Revolution arguments in favor of developing and extending the power of standard French were made in the name of "rationality," during and after that great turning point the standard began to be defended in terms of "nationalism": one national language, one homogeneous identity for all citizens, one set of linguistic resources to allow central governments to tap into the reservoirs constituted by their growing populations. I will return to these "nationalist" waves of homogenization which, in the latter part of the millennium, began to affect the linguistic "stuff" that had accumulated not only in Europe, but in many places outside of it.

Arguments and Operators

I have argued that structures as different as sedimentary rock, animal species, and social classes may be viewed as historical products of the same structure-generating processes. (Or more accurately, of different concrete processes embodying the same abstract machine or engineering diagram.) Does language embody an abstract machine too? The accumulations of linguistic materials that are sorted into homogeneous sets and cemented together through isolation are examples of stratified systems, and, hence, language can be said to embody this (double-articulation) abstract machine. Similarly, insofar as the sounds, words, and constructions of a language are viewed as replicators, languages also embody an abstract probe head, or searching device. But the question we must address now is this: Is there an abstract machine that is specific to language? In other words, do the processes responsible for the generation of phrases and sentences embody an engineering diagram that distinguishes the structure of language from the structure of rocks, plants, and animals?

Chomsky believes that this diagram defines an *abstract robot* embodied in our brains, an automaton capable of producing every valid sentence in a given language. In 1959, Chomsky postulated the existence of four different types of abstract automata which differ in their degree of complexity: finite-state automata are the simplest type, followed by context-sensitive robots, context-free robots, and finally Turing machines.[69]

Chomsky argued that a language could be seen as made up of two components, a dictionary (or reservoir of words) and a *set of rules* determining how those words may be combined to make legal sequences (well-formed sentences). Thus, given a set of sentences, the robot (a context-free automaton) could tell whether they belonged to a given language simply by applying the rules. To the robot, a sentence was no more than a *string of inscriptions* (whether the inscriptions were on clay, paper, or air was immaterial to it), and the rules were recipes to test these strings for membership in the set of valid strings. This model was supposed to capture the grammatical intuition that allows speakers of English to tell the difference between "Colorless green ideas sleep furiously" and "Sleep green colorless furiously ideas" (one a grammatically valid string, the other invalid), even though both strings are semantically meaningless.

When it came time to *produce* new strings (as opposed to checking them for validity), the rules were divided into two types: one set generated the basic logical skeleton of a sentence (its deep structure), while several other sets transformed this naked sentence, fleshing it out with the materials of a real language. (These two components of a grammar are called "generative" and "transformational," respectively.) The generative component of the automaton was assumed to be inborn and to capture all that is universal about language (that is, all that remains constant across different languages and is unaffected by their particular histories). Could we consider this robot the abstract machine of language? Deleuze and Guattari, among others, answer this question negatively:

> Our criticism of these linguistic models is not that they are too abstract but, on the contrary, that they are not abstract enough, that they do not reach the *abstract machine* that connects language to the semantic and pragmatic contents of statements, to collective assemblages of enunciation, to a whole micropolitics of the social field. . . . [T]here is no language in itself, nor are there any linguistic universals, only a throng of dialects, patois, slangs, and specialized languages. There is no ideal speaker-listener, any more than there is a homogeneous linguistic community. Language is, in Weinreich's words, "an essentially heterogeneous reality." There is no mother tongue, only a power takeover by a dominant language within a political multiplicity.[70]

In essence, what Deleuze and Guattari oppose is the postulation of a "universal core" (or synchronic dimension) of language, since this relegates social processes (such as pidginization, creolization, or standardization) to a secondary role, affecting at most the transformational component of the grammar. What they propose instead is to give historical processes a more fundamental role by modeling the abstract machine of language not as an automatic mechanism embodied in individual brains but as a diagram governing the dynamics of collective human interaction. The main problem to be solved if we are to implement their proposal lies in finding a valid means of transferring the *combinatorial productivity* of the automaton, its ability to produce an infinite number of sentences out of a finite stock of words and combination rules, to the patterns of behavior generated by different social dynamics. One possible solution may be to assume that the postulated grammatical rules do not exist in our brains but are instead embodied in social institutions. The problem with this solution is that, as is well known, human beings do not learn their mother tongue as a set of rules. Indeed, it was the well-documented ability of children to learn language by being exposed to adult conversation (that is, without being explicitly told what the rules are) that motivated the postulation of an inborn automaton in the first place. But if a set of rules is not the source of the combinatorial productivity of language, then what is?

One possible answer is that words carry with them, as part of their meaning, "combinatorial constraints" that allow them to restrict the kinds of words with which they may be combined. That is, in this view individual words carry information about their frequency of co-occurrence with other words, so that, as a given word is added to a sentence, this information exerts demands on the word or kind of word that may occur *next*. (For example, after adding a definite article to a string, the following position is constrained to be occupied by a noun.) Combinatorial productivity would not result from a centralized body of rules, but from a decentralized process in which each word *locally restricts* the speaker's choices at each point in the construction. One version of this alternative way of handling the production of sentences was proposed long ago by the linguist George K. Zipf, who was perhaps the first to study language as "stuff," that is, as a large body of material inscriptions exhibiting certain statistical regularities. Zipf called the tendency of words to occur next to each other their degree of crystallization: "To illustrate the comparative degrees of dependence of words in sentence-structure, let us perform an imaginary experiment. We may take as material a vast number of English sentences, just as they are spoken, say a million of them. Figuratively speaking we shall now dash these sentences on the floor with such force that they will break, and pieces of them will scatter. Of course, some of the words, being more crystallized in arrangement than others, will cohere. Definite and indefinite articles will adhere to their nouns, auxiliaries to their verbs, prepositions to following objects."[71]

The linguist Zellig Harris, who introduced the notion of "transformation" into linguistics in the early 1950s (and so is no stranger to the Chomskyan paradigm), has developed a way to take metaphorical descriptions like this and transform them into a mathematical theory of language that comes very close to the abstract machine we are looking for. According to his theory, the constraints or demands that words place on one another are transmitted as socially obligatory information. ("Information" is being used here in the sense of "physical information," the kind measured in bits, *not* the semantic information used in dictionary definitions.) Harris explicitly develops his model of the soc ial transmission of combinatorial constraints in evolutionary terms, with different constraints (or rather, the sentences constructed with their help) competing for the same "informational niches."[72] He rejects the concept of an unchanging, homogeneous core of language, and therefore his theory allows us to approach the question of dialectal variation (and the essential heterogeneity of language) directly: not only is language in constant change, with the strength of the constraints varying along a continuum from optional to obligatory, but the rates of change themselves may be different from dialect to dialect. His view of language is completely historical; the source of the constraints themselves is the gradual standardization (or conventionalization) of customary usage. Thus, despite the fact that changes in syntax may occur much more slowly than changes in other aspects of language, the syntactical element is not isolated from semantics and pragmatics.[73]

Harris classifies three main types of combinatorial constraints. The simplest one is what he calls "likelihood constraints," information carried by words about the words with which they tend to combine *more frequently* as a matter of actual usage. That is, a word like "tiger" carries information to the effect that it typically co-occurs with other words (such as "fierce" or "hunting") but not others ("polite" or "dancing"). Not that there is a specific rule barring these combinations; rather, as a matter of *statistical fact*, in a given speech community these words occur in certain combinations much more frequently than in others. (The phrase "dancing tigers" does occur in children's books, but compared with the overall usage of those two words in actual speech, this combination is rare.) For a given word, the set of its most frequently co-occurring words (a fuzzy set since it is in constant change, contracting and expanding) is called its "selection," and in Harris's model it is this selection set that forms the "core meaning" of the word. (Hence, the meaning of words would be determined by their combinability, not their identity. Formal dictionary definitions and informal stereotypes emerge from conventionalization of likelihood constraints.)[74]

A second type of constraint, the most fundamental to the structure of language, according to Harris, is the operator-argument constraint, which models the action that verbs, adverbs, adjectives, prepositions, and other linguistic modifiers have on their objects. Unlike the likelihood constraint, the operator-argument constraint binds together not individual words but classes of words. A given operator, once included in a sentence, *demands* an argument of a certain class. This constraint, too, adds information to the sentence: the more unfamiliar the argument supplied for a given operator, the more informative it will be. Of all the different linguistic functions that this constraint may be used to model, Harris stresses the operation that verbs perform on the nouns that serve as their subjects and objects, since this operation yields the basic structure of sentences. As is well known, sentences afford their users the means to perform two different functions: to *identify* for an audience the objects or events to which the speaker is referring and to say something *about* those objects or events. The operator-argument constraint, when used to link verbs and nouns, adds to a sentence the meaning of "aboutness," the ability to refer not only to individual objects and events but also to complex *situations*.[75]

Finally, Harris postulates a third type of constraint, which he calls "reduction." Whenever the likelihood that two words will co-occur becomes very high, the amount of physical information their co-occurrence adds to a sentence is correspondingly low; that is, it adds very little information that cannot be supplied by the speaker or listener. In those conditions, one of the two words may be reduced in form (becoming a suffix or prefix attached to the other word) or even eliminated altogether. However, even when a word has been "zeroed," the little information it used to carry is still there (or may be reconstructed by the speaker or listener), so that after successive reductions the resultant simpler forms may carry (in a very compressed way) a rather complex meaning. Harris uses this third kind of constraint to explain the origin of some classes of words (such as adverbs, pronouns, and some conjunctions) as well as of the different affixes.[76] In other words, the reduction constraint allows Harris to give a historical account of the origin of the main word classes, classes which are taken as given (as unexplained primitives) in the Chomskyan theory.[77]

This is one of the reasons why Deleuze and Guattari view the Chomskyan automaton as "not abstract enough." The robot is capable of explaining the production of one set of forms (those of sentences) but only by assuming another set of forms (those of rules and primitive word classes). In Harris's model, on the other hand, language is a thoroughly historical product (the cumulative result of restrictions in the occurrence of words relative to one another), and combinatorial constraints are truly morphogenetic: as new constraints emerge from conventionalization of customary usage, changing the probabilities that words will co-occur, language structure self-organizes as a process involving *successive departures from equiprobability* (i.e., randomness) in the combinations formed by replicating norms.[78]

This scenario meshes well with some of the ideas we developed earlier. In particular, the emergence of language may now be seen as the result of a double articulation: an accumulation formed by a sorting device consolidated through an act (or succession of acts) of conventionalization or institutionalization. However, this diagram may be too simple even to account for sedimentary rocks, which also grow and develop through *accretion*, that is, the amassing of further materials and the proliferation of existing structure. Language, too, in Harris's view, is an accretionary structure.[79] In particular, once certain high frequency co-occurrences have become obligatory constraints, speakers begin to construct new patterns *by analogy* to previously institutionalized ones. Prior structures could also proliferate by *recursion*: operator-argument pairs, for example, themselves could be made the argument of a higher-level operator. Hence, positive-feedback loops develop where structure (consolidated accumulations) favors accretions, which in turn generate further structure. Moreover, the creation of new patterns by analogy to previously accumulated ones (or by recursive application of existing constraints) is what generates a system that, in retrospect, may appear to consist of a set of rules.[80] (Of course, some languages, such as standard English or French, *are* sets of rules, and they are taught to grammar school children as such. The question is whether the language that those children learn at home in an untutored way is also a set of rules or rather a set of normative combinatorial constraints.)

Another feature of Harris's theory may help us meet Deleuze and Guattari's demand that the abstract diagram be "abstract enough." Ideally, the abstract machine postulated to account for the generation of linguistic forms should not be the abstract machine *of language* (in which case it would be hard to distinguish it from an "essence" of language), much as the abstract probe head we discussed before is not the abstract machine *of life* (since it may be "incarnated" in any population of replicators, not only genes). Similarly, an "abstract enough" diagram that explains the generation of strings of linguistic inscriptions should ideally explain the morphogenesis of other (nonlinguistic) strings. In other words, language may not be the only structure that can be viewed as a system of demands or of required repetitions. While the structure of language is unique, the constraints that generate it are not. (Being the subject of a verb is uniquely linguistic; having the occurrence of certain things depend on the occurrence of other classes of things, is not.)

Harris shows how by making the combinatorial constraints more rigid we can generate strings of inscriptions like those belonging to systems of logic or mathematics, while by making them more flexible we can produce musical strings. For example, weak conversational (or discursive) demands constrain the successive order of sentences in ordinary language. If we strengthen those demands, so that sentences must now follow one another in a prescribed manner (and if we further demand that the sequence begin with self-evident truths and conclude with a sentence as true as the previous ones), the result is a logical or mathematical proof structure. If we change the operator-argument hierarchical constraint and demand that only the operator carry constraint-based information, we thereby transform the argument into a *variable* and the operator into a *function*. (That arguments in mathematics exercise no constraints is what makes it a science of relations, that is, of operators.)[81] On the other hand, if instead of fixing the operator-argument relation we make it variable, so that "many varied relations exist between a longer musical line and its subsegments," we can generate structures like those exhibited in musical compositions.[82] This is not to deny that explicit rules exist in mathematical or musical systems, much as they do in standardized languages. The question is whether mathematics or music could have originally developed as a decentralized system of constraints that only later was formalized as a centralized body of rules.

In addition to providing us with an "abstract enough" diagram of language, Harris's theory also meets the other requirement we found lacking in Chomsky's robot: that the abstract machine be directly connected to a social dynamics. Specifically, the core of Harris's model involves a process through which statistical regularities in usage are gradually transformed through standardization into required constraints. But these institutional requirements would have no reality if there was no mechanism through which social obligations could be enforced. It may be argued that to be complete Harris's theory demands some kind of norm-enforcement mechanism, such as that provided by social networks. We saw before that, in sociolinguistics, the degree of density of a network (roughly, the degree to which, for every member of a community, the friends of his or her friends know each other) and its degree of multiplexity (the degree to which his or her life-support activities depend on those friends and friends of friends) are viewed as the parameters that define its efficacy as a norm-enforcement device. In a sense, these parameters define the *intensity of our attachment* to a given community or group, and the norms enforced within a network draw the boundaries that define the identity of that community or group. Thus, a view of language in terms of constraints on word combination directly involves questions of the effects that group-membership has on individuals, and, in that sense, it meets Deleuze and Guattari's requirement that "collective assemblages of enunciation" be made an intrinsic component of the abstract machine of language.

Is it possible to extend (or complement) Harris's model so that a similar abstract diagram explains not only the form and function of individual sentences but also the historical origin of larger linguistic structures, such as discourses? Or more specifically, is there an abstract machine that can explain in sociodynamical terms the emergence of discourses expressing worldviews (coherent sets of values and beliefs)? A model created by the anthropologist Mary Douglas comes close to defining such an abstract machine, and it may be linked with Harris's theory of language since in Douglas's model the intensity with which individuals are attached to a group also defines an important feature of "collective assemblages." Another equally important trait of group dynamics defines not whom we interact with, but *how* we interact; it does not bestow group-membership but controls behavior in the wider social context within which the group functions. Douglas, who calls these two aspects of social dynamics "group" and "grid," one measuring the intensity of group allegiance, the other the intensity of centralized regulation, has created a theory of the self-organization of worldviews, in which the kind of cosmologies that emerge in different communities depend directly on the values of the "group" and "grid" parameters. When applied to specific social groups (Douglas's model does not apply to entire societies), these two parameters define an abstract diagram with four possible stable states that act as "attractors" for beliefs and values as they organize into a coherent set. Or rather (since she models not the dynamics of beliefs but the dynamics of *groups of believers*), the two parameters define a lifestyle (more or less hierarchical, more or less group-dependent) and people *coerce one another* to fully develop the implications of that lifestyle. The resultant worldviews act as attractors in the sense that "the four extreme grid/group positions on the diagram are liable to be stable states, steadily recruiting members to their way of life, which is at the same time inevitably a way of thought."[83]

When both the group and grid parameters have high values, the community in question not only has a strong sense of self-identity (the group may spend much energy policing boundaries and elaborating rules of admission) but it is also well integrated into larger social groups. Life within a government military institution such as the army or navy would serve as a good example of this lifestyle, but so would the culture of any hierarchical bureaucracy. Keeping the value of group allegiance high but lowering the value of regulation (and integration into a larger whole) results in sectarian lifestyles with strong group identity but a weak sense of responsibility to conform to any norms that hold outside the group. If both parameters are set at a low intensity, group members refrain from drawing strong boundaries around them (they rather engage in networking; given the loose group demands, everything seems open for negotiation), and they tend to participate in those areas of public life that are less centralized and hierarchical. (A small-business entrepreneur would be a good example here, but not the manager of a large corporation, particularly if he or she participates in the corporate culture.) Finally, there are those who do not belong to closed groups but nevertheless have little room to maneuver around regulations and are, indeed, burdened by them:

As I see it, three corners exert a magnetic pull away from the middle; individualists extolling a culture of individualism tend to become more and more uncommitted to each other and more committed to the exciting gamble for big prizes. Egalitarian idealists committed to a sectarian culture strongly walled against the exterior, become more and more enraged against the outside society and more jealous of each other. The supportive framework and intellectual coherence of a hierarchical and compartmentalized society nurses the mind in cogent metaphysical speculations vulnerable to disorder and independence.... The fourth corner, the fully regulated individuals unaffiliated to any group, is plentifully inhabited in any complex society, but not necessarily by people who have chosen to be there. The groups [bureaucracies or sects] expel and downgrade dissenters; the competition of individualists ... pushes those who are weak into the more regulated areas where their options are restricted and they end by doing what they are told.[84]

Although Douglas's model may have to be enriched in several ways, even in this simple form (with two parameters generating four possible states) it meshes well with the ideas we have explored in this book. First of all, it attempts to capture some of the features of group dynamics behind the *genesis of form* at the level of coherent discourse. That this morphogenetic process may turn out to be more complex does not deprive her hypothetical model of validity as a first approximation, particularly if the model is given a nonlinear dynamic formulation so that the first three corners of the diagram become true attractors. (A catastrophe theory version of Douglas's model does exist and points in the direction that this reformulation would have to take.[85]) Additionally, the model is intended to be used in a bottom-up way, to be applied to the study of specific communities, where the constraints that the holders of a worldview exert on one another can be fully specified. In other words, the scheme is not supposed to apply to societies as a whole but only to smaller subsets thereof, with cities or nation-states modeled as complex mixtures of several types of worldview.[86] On the other hand, Douglas's model has limitations: it only captures processes that take place *within* organizations or collectivities, and hence cannot account for the effects of the transmission of ideas and routines between the members of an ecology of institutions or, indeed, for any effect on the form of discourses which the interactions between institutions may have (e.g., the interactions between hospitals, schools, prisons, and factories).

Returning to the question of the abstract machine of language, both Harris and Douglas have contributed crucial insights into the essentially *collective character* of this machine. In both linguistic evolution and world-view development there are, no doubt, many contributions and innova-tions by individuals. But in many cases it is the position of an individual in a communication network that determines the fate of his or her contri-bution. Consequently, the accumulation and consolidation of languages and worldviews is a collective enterprise, not the result of individual self-expression. Moreover, to the extent that the resulting linguistic and dis-cursive forms are transmitted to new generations (or new members) through enforced repetition, these forms are replicators; hence we need to use "population thinking" to describe their evolutionary dynamics. This, too, forces on us the need to approach the subject in terms of col-lectivities rather than individuals. On the other hand, the collective dynamics may be such (low group/low grid) that individuals may play sig-nificant roles in the fate of these accumulations. But even so, it may be argued that this extra room to maneuver is afforded to individuals by the stable state governing the collective dynamics, and in any case those indi-viduals owe their surplus freedom to the fact that they are connected to decentralized structures (such as markets), which are every bit as collec-tive as the most routinized hierarchy.[87]

We may now picture the structure-generating processes behind individ-ual sentences as embodying an abstract machine operating on the basis of combinatorial constraints transmitted as replicators. The process of transmission itself involves collective mechanisms of enforcement, which are also part of the abstract machine of language and which may be used to account for the emergence of coherent structures made out of many sentences (discourses embodying specific worldviews). Now we must return to the historical development of both these components of the abstract machine and examine the history of their multiple and complex interactions.

Linguistic History: 1700–2000 A.D.

The eighteenth century witnessed two dramatic developments that were to affect profoundly the fate of the linguistic mixtures of Europe: the rise of nationalism and the growth and spread of disciplinary institutions. On one hand, the project of nation building was an integrative movement, forging bonds that went beyond the primordial ties of family and locality, linking urban and rural populations under a new social contract. On the other hand, complementing this process of *unification*, there was the less conscious project of *uniformation*, of submitting the new population of free citizens to intense and continuous training, testing, and observation to yield a more or less uniform mass of obedient individuals. This was not to be, however, an undifferentiated mass, but one whose components were sufficiently individualized to then be sorted out into the ranks of the new meritocracies, where achieved status replaced ascribed status as the criterion for establishing rank. Although unification and uniformation did not always go together (and, indeed, their requirements sometimes clashed[88]), certain countries underwent both processes simultaneously. As Foucault writes:

> Historians of ideas usually attribute the dream of a perfect society to the philosophers and jurists of the eighteenth century; but there was also a military dream of society; its fundamental reference was not to the state of nature, but to the meticulously subordinated cogs of a

machine, not to the primal social contract, but to permanent coercions, not to fundamental rights, but to indefinitely progressive forms of training, not to the general will but to automatic docility.... The Napoleonic régime was not far off and with it the form of state that was to survive it and, we must not forget, the foundations of which were laid not only by jurists, but also by soldiers, not only councillors of state, but also junior officers, not only the men of the courts, but also the men of the camps. The Roman reference that accompanied this formation certainly bears with it this double index: citizens and legionaries, law and manoeuvres. While jurists or philosophers were seeking in the pact a primal model for the construction or reconstruction of the social body, the soldiers and with them the technicians of discipline were elaborating procedures for the individual and collective coercion of bodies.[88]

In France, not only did unification and uniformation coexist, but they reached a higher peak of intensity during the Revolution of 1789 than in other European nations. In particular, the revolutionary armies, later to become the core of the Napoleonic war machine, were the perfect embodiment of both projects. These were citizen armies, unlike the mostly mercenary armies that had heretofore dominated European warfare, and therefore larger and stronger in morale. They were a manifestation of the fact that the new social pact had transformed the growing population of France into a vast reservoir of manpower, to be tapped not only for political participation in the new democratic institutions, but also as a massive source of new recruits. In order to function as part of a larger machine, however, these masses would need to be "processed" by means of novel methods of drill and exercise and continuous observation and examination, which alone could transform these human raw materials already possessed of nationalistic fervor into efficient components of a new combinatorial calculus in the battlefield (e.g., the tactical system of Jacques-Antoine de Guibert).[90]

Both the meshwork of dialects and the superimposed hierarchical standard languages were affected in a variety of ways by these two social projects. Around 1760 (in France as well as in other countries), dialectal variation came to be seen not as a question of inferior rationality relative to the standards, but as a problem of the state: an obstacle to unification and national consolidation, a potential source of local resistance to integration into the larger social body. During the French Revolution, this new attitude toward language led to intolerance, not only toward aristocratic Latin, but also toward the dialects and patois (dialects without a writing system) that the majority of French citizens spoke, but which now represented provincialism and backwardness to the Parisian elites. This linguistic chauvinism was expressed thus by a revolutionary in 1794: "Federalism and superstition speak Breton; emigration and hatred of the Republic speak German; the counterrevolution speaks Italian, and fanaticism speaks Basque."[91] During these turbulent years, speaking French came to be seen as a political act, an expression of patriotism. Revolutionaries were divided as to what counted as "politically correct" French (the sansculottes wanted it "brutal and vulgar," while the leaders of the French national assembly preferred it "free, bold and manly"), but they were united in their common distrust of the dialects, which they believed the enemy might use to fragment and marginalize the masses.[92]

The study of Greek and Latin in school was viewed by revolutionaries as a transmission of dead knowledge through dead languages and was eventually forbidden. (Napoleon, a classicist himself, would reinstate the instruction of Latin in schools later on.) The counterrevolution, on the other hand, defended Latin on the grounds that it allowed the dead to speak to the living, thus providing continuity with the classical past, a continuity threatened by the rising vernaculars. Behind this struggle over the relative merits of major languages (both sides despised minor languages), there was a philosophical attitude toward language in general that found expression in these years and laid the foundation of the dispute. The idea that the structure of language determines the structure of perception may have originated with Diderot and Condillac, and it first acquired political overtones during the French Revolution. Both sides took this idea very seriously. The revolutionaries found covert, oppressive meanings hiding behind old words (especially behind aristocratic titles and names) and added to their political agenda the "relanguaging" of everything, from the French calendar to place-names. The counterrevolution, on its side, saw its enemies as "drunk on syllables; rioting in an orgy of words, issuing from suffocating rivers of speeches, books and pamphlets."[93] A fundamental misunderstanding of the meaning of certain words (e.g., *egalité, volonté*), the royalists believed, had shaped the revolutionaries' thought processes and distorted their perception of things.

Whatever the merits of this view of the nature of language and perception, a national language was felt necessary because only through linguistic unity could the emerging elites mobilize the masses for peace and for war. A uniform means of communication was needed to transmit the new political ideals to the people and allow their participation in a national political process. It was also necessary as a means of exhortation (to tap into the human reservoir by manipulating nationalistic feelings) and as an instrument of command in the army. This latter task became even more important as Napoleon transformed warfare from the dynastic duels typical of the eighteenth century to the kind of "total war" with which we are familiar today, a form of warfare involving the complete mobilization of a nation's resources. In this regard, one of the most important "innovations" of the Revolution was the creation of a recruitment system that amounted to universal conscription or compulsory military service.[94] The transformation of the French population into a human reservoir to be mobilized for total war was initiated by an institutional speech act, a decree issued by the National Convention in August 1793:

> ...all Frenchmen are permanently requisitioned for service into the armies. Young men will go forth to battle; married men will forge weapons and transport munitions; women will make tents and clothing and serve in hospitals; children will make lint from old linen; and old men will be brought to the public squares to arouse the courage of the soldiers, while preaching the unity of the Republic and hatred against kings.[95]

Of course, as with all speech acts, this decree's power to catalyze a major social change depended on many nonlinguistic factors, such as the existence of a growing urban population without clear economic prospects and an administrative apparatus capable of handling the bureaucratic tasks demanded by such a massive mobilization.[96] The efficacy of the decree also depended on an intensification of the uses of discipline, supervision, and examination. A similar remark can be made about the institutional speech acts that abolished the use of Latin and non-Parisian dialects during the Revolution. In particular, the "Frenchification" of the provinces was not a project that could be realistically carried out at the end of the eighteenth century, because there was yet an insufficient number of teachers. (This process would have to wait about a hundred years, until 1881–1884, when primary education in standard French was made compulsory.[97]) Additionally, schools had to be transformed into disciplinary institutions, a slow process that had begun before the Revolution. Throughout the eighteenth and nineteenth centuries, schools evolved within a complex institutional ecology (that included hospitals and barracks, prisons and factories), increasing their use of writing to record individual differences, of repetitive exercises for both training and punishment, and of a system of command based on signals that triggered instant obedience. As Foucault observes, "The training of school-children was to be carried out in the same way [as in the army]: few words, no explanation, a total silence interrupted only by signals—bells, clapping of hands, gestures, a mere glance from the teacher."[98]

One should be careful, however, about extrapolating Foucault's findings to other countries, because eighteenth-century France was a pioneer in this regard. Her arsenals and armories were at this time developing one of the key elements of mass production; her language academy was the world's leading standardizing institution; and, finally, most other nations implemented democratic institutions and replaced their aristocracies with meritocracies without painful revolutions and over much longer periods of time. England (where these changes were effected only after seven decades of social reform, 1832–1902) is illustrative here precisely because it involved such different conditions. In particular, a key element of the process of nation building—one that France was late in implementing— was the creation of a nationwide market. As we observed in the first chapter, unlike local and even regional markets, national markets were not self-organized meshworks but involved a good deal of command elements emanating from the capital city. If Paris played the role of intellectual hothouse, where the ideas and energy behind the Revolution accumulated and synthesized, London played the role of a huge economic machine animating trade flows throughout England. Both capitals were ultimately parasitic, and yet they were also essential to the process of forging a uni- fied, hierarchical national entity out of a meshwork of provinces and regions:

> These towns . . . represented enormous expenditure. Their economy was only balanced by outside resources; others had to pay for their luxury. What use were they therefore, in the West, where they sprang up and asserted themselves so powerfully? The answer is that they produced the modern states, an enormous task requiring an enormous effort. They produced the national markets, without which the modern state would be a pure fiction. For, in fact, the British market was not born solely of the political union of England with Scotland (1707), or the Act of Union with Ireland (1801), or because of the abolition of so many tolls . . . or because of the speeding up of transport. . . . It was primarily the result of the ebb and flow of merchan- dise to and from London, an enormous demanding central nervous system which caused everything to move to its own rhythm, overturned everything and quelled everything.[99]

Here, too, we find the same combination of institutional speech acts instantly creating political unions or destroying economic obstacles (tolls), and an energetic and material process (intensified trade flows) sustaining the efficiency of those linguistic catalysts. The most important form of merchandise flowing from London in the eighteenth century, in terms of its effect on linguistic materials, were the "linguistic engineering" devices constituted by authoritative (and authoritarian) dictionaries, grammars, and guides to proper pronunciation. Unlike in France, these would not be the product of government institutions (academies) but of individuals tak- ing advantage of the emerging national market, which amplified their efforts as much, or more, than any nationwide organization could. These devices, perhaps best illustrated by Samuel Johnson's dictionary of 1755, had a long-lasting effect on the English soup of linguistic replicators, increasing its homogenization and the subordination of all other dialects to the written standard of London. The social dynamic of London and other large towns, where the middle class was growing in number and importance, greatly facilitated the penetration of these devices, since, as we observed earlier, it is in socially mobile classes that the pressures from social networks to preserve local linguistic patterns as badges of identity are at their weakest.

When Dr. Johnson published the first edition of his dictionary, London had already experienced a whole century of authoritarian attitudes toward language, mostly inspired by writers such as John Dryden, Daniel Defoe, and Jonathan Swift. These writers publicly decried the "corruption" of the English language by spontaneous linguistic change and lamented the lack of an academy on the French model to protect the "purity" of the language by fixing it in its pure state for all time. (Defoe, for instance, wanted to make the coining of new words as criminal as coining money.)[100] But nothing came of these calls for linguistic reform until Dr. Johnson's dictionary codified the lexical features of English, that is, recorded "reputable" vocabulary and exhibited "correct" pronunciation patterns:

> So strong was the social influence of Dr. Johnson that his work became synonymous with the word *dictionary* itself, and *the* dictionary dominated English letters for over a century and remained in use until 1900. One measure of the dictatorial power of "the Dictionary" is the fact that a Bill was thrown out of Parliament in 1880 simply because one of its words had not been recorded by Dr. Johnson. This mystical power soon extended to other dictionaries in the latter half of the eighteenth century, especially with regard to proper pronunciation. Speakers of middle-class dialect, eagerly engaged in social climbing, wanted authoritative keys to the articulations of polite society. As a result of this ready-made market, pronunciation dictionaries thrived during the last three decades of the eighteenth century.[101]

A few years after Dr. Johnson's dictionary was unleashed on the population of linguistic replicators, decreasing the intensity of their variation, a series of normative and prescriptive grammars began to be published with the aim of reducing the syntactic habits of London's upper classes to a set of codified principles. Although by today's standards their efforts were not scientific (they used synthetic Latin grammar as a kind of "universal grammar" to codify English, which had already become an analytic language), the early grammars had a great impact in their time and many of their prescriptions and proscriptions (e.g., discouraging the ending of sentences with prepositions and the splitting of infinitives) are still with us today.[102] Together with dictionaries, these mechanically reproduced sets of norms furthered the London standard's domination of other dialects. However, much as standard French would need to wait for compulsory primary school to become a true homogenizing force, so would standard English remain a coexisting (if more prestigious) norm until 1870, when primary schooling was declared "universal" and children began to learn English twice: once as a living language at home and again as a set of codified rules at school.

Thus, in the eighteenth and nineteenth centuries, standard French and English continued to widen their power base at home. They also began, via colonialism and conquest, to spread around the world. At this point, despite the growing size and power of the British Empire, English was still inferior to French (and even to Italian and Spanish) in terms of international prestige. But this would soon change, and during the following century the number of English speakers in the world would rise sharply (almost tripling between 1868 and 1912), as would its rank in the global pyramid of colonialist standards.[103] Eventually (in our own century), English would challenge French for the role of "world standard." But even before the twentieth century, the colonial competition among the European powers—and the concomitant spread of their languages throughout the world—was already an important element of a process that would eventually lead to global confrontation.

Western colonialism was reproducing, on a worldwide scale, the conditions in which Europe found itself at the turn of the millennium. Instead of one imperial standard (written Latin) immersed in a complex mixture of vernaculars, now a variety of standards (first Spanish, Portuguese, and Dutch and later on English and French) coexisted and interacted with an even more varied combination of local languages. The situation was not, of course, exactly analogous since the soup of linguistic materials surrounding written Latin was largely made up of divergent forms of spoken Latin, while in the centuries between 1500 and 1900 European languages came into contact with populations of norms which had been shaped and sculpted by distinct and diverse historical forces. Furthermore, the number of different contact situations that were created during these centuries exceeded those that existed when the Romance languages were forming. Thus, while commercial contacts in both periods produced trade pidgins (Mediterranean Sabir and Chinese pidgin English, respectively), only the second period produced situations where new stable languages could crystallize. Indeed, as Dell Hymes has said of modern pidgins and creoles:

> Their very existence is largely due to the processes—discovery, exploration, trade, conquest, slavery, migration, colonialism, nationalism—that have brought the peoples of Europe and the peoples of the rest of the world to share a common destiny. More than any variety of language, they have been part of these activities and transformations.... And while these languages have come into being and existed largely at the margins of historical consciousness—on trading ships, on plantations, in mines and colonial armies, often under the most limiting or harshest of conditions—their very

origin and development under such conditions attests to fundamental characteristics of language and human nature.[104]

Slave plantations are perhaps unique among the different contact situations generated by the expansion of Europe. Plantations became veritable "linguistic laboratories" where brand-new languages were produced
out of elements of African dialects and a streamlined version of a major
European language. As we argued above, far from being "corruptions" of
the master's language, pidgins must be viewed as creative adaptations
developed by the slaves themselves in order to communicate with each
other. As one linguist points out, "All the early accounts (dating from the
eighteenth century in Jamaica, for example) report that the white plant-
ers and their families were learning the creole from the slaves, not vice
versa."[105] Slaves needed to invent their own lingua franca because plantation owners deliberately purchased Africans with different linguistic backgrounds to prevent them from communicating with one another, hence
reducing the risk of insurrection.

So far I have been using the term *pidginization* to refer to any process
of reduction or simplification of linguistic resources, including the conversion of a synthetic into an analytic language. Although there are linguists
who use the term in this sense (e.g., William Samarin[106]), Hymes has
objected that simplification alone cannot account for the birth of (more or
less) stable entities, such as the precursors of Jamaican English and Haitian French. Hymes adds the requirements that the new, simplified pidgin
be used by several groups (each with its own mother tongue distinct from
the pidgin) and that there be an admixture of linguistic materials from
different sources. To this it should be added that the language being
pidginized—in the case of plantations, the master's language—must be
absent both as a source of stigmatization and as a reference model. That
is, the crystallization of a pidgin involves a barrier (geographical or social)
that distances the emerging entity from the conservative tendencies of
the prestigious target language. Only under these conditions can a pidgin
achieve autonomy from the dominant norm, and it is this autonomy that
defines it as a separate entity.[107]

Another difference between the pidgins generated by European colonialism and those that emerged (before and after) as trade jargons is that
the plantation pidgins, after their slave speakers became free, soon
evolved into more durable entities called "creoles." One way in which pidgins avoid extinction is precisely by reenriching themselves with many of
the redundant features eliminated during the simplification process and
by diversifying in the number and type of uses they can be put to. Accord-

ing to one important theory of creolization, recognizing that many plantation creoles are a one-generation process, children play a crucial role in a creole's recomplexification. Children's ability to do this may be explained as deriving from internal linguistic structures (i.e., Chomsky's robot) that are universal to all languages and expressed most fully in the critical years of childhood when language acquisition is easier. (This is the current explanation for the creolization of Hawaiian pidgin, for example.)[108]

On the other hand, the role of children in the creolization of plantation pidgins may be explained in terms of sociolinguistic constraints. Given that adults who have just undergone the transition from slavery to freedom cannot be expected to feel a great deal of loyalty to their pidgin (which was not a badge of local identity), they do not behave toward it as a traditional norm to be preserved. Therefore, as they transmit these norms to their offspring they exert very little effort to suppress novel utterances, so that many nonstandard words or phrases survive and are eventually used to reenrich the pidgin.[109] As usual, we may expect complex and varying mixtures of these and other factors to be responsible for specific creoles around the world. More importantly, varying mixtures of factors will be active in different regions of the same country, as in the separate plantations of Jamaica. When one speaks of the crystallization of a creole (or a pidgin) as a separate entity, one must also keep in mind that these novel entities are still part of a continuum of dialects, much as nonstandard English or French are in their home countries. Therefore, to speak of Jamaican or Haitian creole is to refer to that segment of a continuum of variation which exhibits the maximum divergence from the standard but which is still connected to other portions of the meshwork.[110]

Today, the majority of creole speakers live in the Caribbean Islands (about six million), although there are also smaller populations in western and southern Africa and southern and southeast Asia. The Caribbean is numerically dominated by French-based creoles, but a million and a half Jamaican creole speakers speak an English-based dialect. The absence of Spanish and Portuguese creoles in this region is puzzling, given that they are widely spoken in Asia and that Spain and Portugal's presence in the Caribbean antedates by more than a century the arrival of the French and British. (Papiamento is the only example of a Spanish-Portuguese creole, but it incorporates so many Dutch and English elements that it is almost a creolized Esperanto.)[111]

The historian Sidney W. Mintz offers one explanation for this apparent anomaly in terms of the demographic and social conditions that sustained the special contact situation outlined above. As he says, plantations were not real communities but socially artificial collocations of slaves and masters the political basis of which was raw physical force. In the Caribbean, plantations were part of a repeopling of "population vacuums" created by European weapons and diseases. All plantations in America had this in common. But there were differences as well: "Generally speaking, the Hispano-Caribbean colonies were never dominated demographically by inhabitants of African origin; moreover, in those colonies movement from the social category of 'slaves' to that of 'freemen' was almost always *relatively* rapid and *relatively* continuous."[112] By the time the number of African slaves increased significantly in Spain's colonies (late-eighteenth-century Cuba and Puerto Rico), the islands had already ceased to be population vacuums and were now peopled by speakers of Spanish.

These differences (demographic composition and degree of social mobility) directly affected the conditions under which stable entities arose. The more numerous the slave population relative to the masters and the slower the "phase transition" from slavery to freedom, the more distant and inaccessible the dominant linguistic norm would be for the slaves, a circumstance that promoted the autonomy of the pidgins and creoles. Other circumstances were also "barriers" to the norm, such as the attitude of the white colonists toward their homeland. "Whereas the Spanish settlers in Cuba and Puerto Rico soon came to view themselves as Cubans or Puerto Ricans, the French and British colonists apparently tended to see themselves as Europeans in temporary exile."[113] One factor affecting this attitude was the rigidity of administrative control exercised by the capitals and metropolises of Europe: the more rigid and uncompromising the colonial policy, the easier for the colonists to establish a local identity. This in turn may have affected other factors, such as the growth of an intermediate mulatto class, which depended on the readiness of the colonists to mix racially (highest among the Spanish, lowest among the British, with the French in an intermediate position). These intermediate classes (and their limited but real social mobility) affected the sociolinguistic situation, decreasing the focus of the transmission of linguistic replicators and hence the ease with which the emergent norms could become autonomous.

In summary, while the dialects of Paris and London were being artificially frozen through standardization in their home countries, elsewhere their constitutive norms were being operated on by those under Europe's colonial rule, producing the opposite result. That is, while academies (or the combination of national markets and linguistic engineering devices) were consolidating a pyramid of dialects in Europe, the major European languages at the top of those hierarchies were being resculpted and adapted for different purposes by minorities around the world, resulting in a continuum of variation of which the crystallized creoles represented only one (maximally divergent) segment. As we move on into the nineteenth century, other contact situations created mixtures of factors and interactions between local and European languages which resulted in different appropriations of English and French. During the nineteenth century, the continent that underwent the most intense form of colonialism was Africa, which was carved up between Britain, France, Germany, and other European powers. These countries assumed control of different regions, most of which were linguistically heterogeneous, enclosed them within arbitrary borders (that is, frontiers cutting across preexisting ethnic and tribal boundaries), and imposed their language as the official tongue of colonial administration.

Much as differing attitudes toward administrative policy resulted in different linguistic outcomes in the case of plantation creoles, so, too, in the conquest of Africa: England (and Germany) followed a policy of "indirect rule," according to which existing institutions were allowed to survive and were used to govern the colony; France, on the other hand, was more inclined to export her own institutions into her colonies. These different attitudes were also reflected in the (explicit or implicit) linguistic policies of the conquering powers. The French projected their language (which they believed to embody universal values of clarity and rationality) with missionary zeal, while the Germans were contemptuous that "lesser breeds" would express themselves in German and therefore did not export their language to the colonies. England was intermediate between the two, not actively promoting English but willing to bestow it on the elites of the regions under her rule. For the same reasons, the French emphasized assimilation and hence were much less tolerant of local languages (and culture), while the British and Germans stressed social distance and allowed their languages to coexist with local varieties.[114]

The main difference between the linguistic contact situations that arose in nineteenth-century Africa and those that occurred earlier on Caribbean slave plantations is that the former did not occur in a populational (and therefore linguistic) vacuum, but rather involved a coexistence of different peoples and languages. In particular, the conquering linguistic norms from Europe faced three strong rivals in Africa: Arabic (mostly in the north), Hausa (the prestigious language associated with the pomp and splendor of the ruling elite in northern Nigeria), and finally Swahili (a language of creole origins which had by then become a lingua franca on so linguistically heterogeneous a continent). Written Arabic had, at this point, the solidity of a standard language, given the tendency of its users to imitate the language of the Koran, whose every word was supposed to have come directly from the mouth of God. Hausa and Swahili were also "Islamized" to a certain extent, and yet Swahili, due to its role as a lingua franca (and hence the tendency of its constitutive norms to replicate across ethnic and tribal frontiers), was more ecumenical than Hausa or Arabic.[115]

From the perspective of the conquering powers there were two reasons to get involved in linguistic matters. On one hand, government institutions were interested in tapping into the reservoir of African peoples for menial clerical positions. (Later on, during the two world wars, their interests would shift to converting this reservoir into a source of recruits for Western armies.) The colonial governments needed, therefore, a language of administration as well as a language of command. On the other hand, Africa underwent the most intense Christianization of any continent after 1800, a process that involved ecclesiastical institutions (or their missionary representatives) not only in the effort to diffuse their spiritual values among the subject population, but also to spread a Western-style educational system. Here the need was twofold: a common language of instruction (typically a Western one) was necessary, but so too was the elaboration of local languages in order to transform them into vehicles for spiritual communication. (Missionaries, for example, devised orthographies, grammars, and dictionaries for many African vernaculars in order to translate the Bible into them and preach to the locals in their mother tongue.) These two different forms of cultural assimilation often came into conflict: the British and German policy of using existing institutions to govern meant that, wherever those institutions were Islamic, the regions under their control were off-limits to the Christianizers.[116]

Both Britain and Germany picked Swahili (in addition to English or German) as their language of administration and command. Unlike Hausa, which was strongly identified with a specific tribal elite, Swahili was a more ethnically neutral dialect. It was likely the Germans in Tanganyika who gave Swahili the greatest impetus. German missionaries helped codify some of its features and extend its uses. By 1888, newspapers were being published in Swahili. The British, on the other hand, adopted Swahili in a more subordinate role (for instance, for use in the lower courts, while English still dominated the higher courts).[117] The sociolinguistic situation of the different African territories also influenced government policy on language. In Tanganyika, where there was more linguistic fragmentation (there were no large kingdoms around which linguistic loyalties might have coalesced), Swahili seemed to be the only choice. In Kenya, the population was much more concentrated into linguistically homogeneous blocks in the well-watered highlands (each block separated by land ecologically unreceptive to European settlement), so the dominant local dialect, whatever it happened to be, was as good a choice as Swahili as the language of administration. (This ambiguous role would later on have consequences for Swahili. After independence, Tanganyika, now rebaptized as Tanzania, adopted it as a national language. Kenya did not.)[118]

Regardless of these local variations, Swahili was always subordinate to English. Even in postindependence Tanzania (where street signs, coinage, public notices, and town meetings use Swahili), this creole is used only for primary education, while English remains the vehicle for higher education and international communication (hence, it is the language associated with social mobility). Although only a few elites (e.g., in Liberia) speak English as their first language, it has become the most important second language in two-thirds of Africa. Under these circumstances, it has become important for Anglophone Africans to appropriate English for themselves and set it in variation so that it can evolve into a creole uniquely suited to their linguistic needs.[119]

In the linguistic conquest of Africa, English did better than French, which became the second language of only one-third of African speakers. But as important as Africa was in the contest between these two languages, the decisive battles in this rivalry would be fought on other continents. In particular, English became the language of four out of five neo-Europes (though it shared the fourth, Canada, with French). Because of the extreme fertility of these temperate zones, English speakers multiplied at a significantly faster rate than French speakers. As in other colonies, settlers in the United States, Australia, and New Zealand reinjected their colonial language with heterogeneity, as they entered into a number of different contact situations through which linguistic items from foreign tongues penetrated English. Settlers adopted a number of terms, particularly names of places and unfamiliar plants and animals, from Native Americans and Australian Aborigines. Yet, as happened to Celtic in relationship to Old English, the norms borrowed from the subjugated peoples had a very high death rate (e.g., of the 130 terms American English borrowed from the Algonquian family of Indian languages, only a fourth have survived to the present day).[120] Contact with colonists from other countries (France, Spain, and Holland, in the case of the United States) also produced a flow of linguistic loans of varying durability, as did the languages swept in by several waves of immigration. (German seems to have been the first immigrant language to have had a marked influence on American English.)

However, by the mid nineteenth century, technological developments were working against these heterogenizing forces. In particular, the intensification in the speed of local and global communications brought about by steam power (in locomotives and transoceanic ships) and electricity (telegraphs) meant that one indispensable element in the creation of new languages, isolation after contact, was now harder than ever to achieve. As we have observed, the entities that form out of a flow of replicators (whether genes, memes, or norms) that has been sorted by selection pressures need to be isolated from other replicative flows in order to consolidate into a new entity. The barriers that create these isolated pockets of replicators can be of different types. To distance and geographic inaccessibility, we must add the emotional barrier constituted by loyalty to a local variant (in dense social networks), the mechanical barrier of different articulatory systems (hard-to-pronounce foreign words), and even conceptual barriers (words are not readily transferred to or from a language that has no "words" in the Indo-European sense). The linguist Keith Whinnom argues that these four types of obstacles to linguistic diffusion have close counterparts in the case of genetic replicators (ecological, behavioral, mechanical, and genetic barriers).[121]

In the case of American and Commonwealth English, only the first two barriers (distance and loyalty) could have played a role in the generation of new entities, much as they did centuries before when Middle English developed into five distinct dialects. But as ships, trains, and telegraphs began to "shorten" geographical distances, only loyalty to local variants remained as a defense against homogenization. Under these circumstances, American English did not develop its own strongly individuated dialects, but only weakly differentiated "regionalisms."[122] On a more global level, the intensified speed of communications meant that British, American, and Commonwealth English (at least in their standard versions) would from now on tend to converge rather than diverge. In a sense, steam transformed English into a single "norm pool" much as it helped microorganisms form a single disease pool. Alongside this long-term process, however, there were shorter-term processes that reinjected heterogeneity into the different pools of linguistic replicators, taking advantage of the one barrier that had not collapsed under the weight of industrialization: emotional attachment to variants that served as local identity badges.

In the United States there were different versions of this emotional attachment, ranging from the nationalism of Noah Webster, who between 1783 and 1828 published grammars and spellers and the local equivalent of Dr. Johnson's authoritative dictionary, to the emergence of black vernacular English, perhaps through creolization of a plantation pidgin. To this already complex mixture of replicators, the nineteenth century would add yet another element, which had both homogenizing and heterogenizing effects: the first mass medium, the large-circulation newspaper.

Although the one-penny newspaper was born in England in 1816, the tendency of the British government to control the press through taxes made it hard for this new medium to spread in London as fast as it did in New York City, where numerous cheap newspapers began to appear in the 1830s. (Freedom of the press, a principle first codified in the United States Constitution, was partly a response to efforts by the British colonial administration to tax prerevolutionary American newspapers.)[123] In one-penny papers such as the *New York Sun* (1833) or the *New York Herald* (1835), "crime and scandal" journalism first found expression. Given the popular appeal of these themes and of the personalized, sensationalistic style of presentation, these papers were the first to bring *massification* of opinion and commercial advertising together. The principle of freedom of the press was conceived to encourage an older type of newspaper, serving as "the means of communication between the government and important groups in society, or between members of the same groups challenging for political power,"[124] and yet in the end it was the commercial type that came to prevail. (Hence, the principle did not lead to a "free marketplace" of ideas, but to a general contraction of opinion.)

The very idea of massified advertising meant that large-circulation newspapers were *not* in the business of selling information to people, but rather of selling *the attention of their readers* to commercial concerns. I have already mentioned several ways in which language was used in the nineteenth century to tap into the reservoir of resources constituted by the growing urban populations in order to mobilize them for political participation or military service. Mass advertising added yet another way of exploiting this reservoir, by mobilizing their attention. At first, both markets and antimarkets used this new resource, but our experience in the twentieth century clearly indicates that big business was soon to be the main beneficiary of this novel way to tap populational reservoirs.

The new mass medium itself would soon join the ranks of the antimarket. Indeed the only clear tendency that one can discern in its two-hundred-year history is precisely a tendency toward increased concentration of ownership and increased scale of production (both of which threaten the freedom of the press).[125] These tendencies were already discernible in the nineteenth century. On one hand, the production of large-circulation papers depended on access to expensive technology, such as the rotary press (capable of printing twenty thousand papers in one hour), new paper-production techniques (wood pulp replaced rags as the principal raw material by the 1880s), and even page composition via key-boards (the Linotype of the 1890s). This meant that as a business, news-paper publishing became heavily capitalized, which acted as an entry barrier for new entrepreneurs. Also, the first casualties of circulation wars, such as the one fought by Pulitzer and Hearst in the 1890s, were often small newspapers.[126]

Furthermore, some segments of the industry began to engage overtly in anticompetitive practices, such as the formation of a cartel by six New York papers, which resulted in the formation of the Associated Press in the 1860s, a news agency that monopolized access to two of the largest European news agencies, the French Havas and the British Reuters. These two agencies in turn had signed an agreement in 1859 (together with the German news agency Wolff) to carve up the world into spheres of influence, with each agency having a virtual monopoly to sell interna-tional news services to these captive markets. Reuters got the British Empire plus China and Japan. Havas acquired control over the French empire and Spain, Italy, and Latin America, while Wolff monopolized access to Germany, Russia, and Scandinavia.[127] Although the profits that these agencies generated were never great (as compared with other anti-market institutions at the time), the agencies nevertheless accumulated a great deal of power, which they exercised, for example, by protecting their turf from the numerous national news agencies that were develop-ing at the time.

The overall effect of mass newspapers and news agencies was homog-enizing. Newspapers aimed their presentation to the lowest common denominator, while news agencies attempted to create a product that would be acceptable to all their subscribers (i.e., newspapers with vastly different editorial policies), which meant that rather than aiming for objectivity they aimed for widely acceptable neutrality. "The agencies assume that a uniform editorial approach is not only possible but also desirable. A government crisis is covered in the same way whether it hap-pens in Nigeria or Holland. Similar standards are applied whether the story is being sent to Pakistan or Argentina. A single, objectively verifi-able account of each event [which in most cases means quoting a reliable official source] is the bedrock of agency reporting."[128] It is this homoge-nization of point of view, amplified by the news agencies' global reach, that is the real problem with the agencies today, not some overt conspir-acy to diffuse "capitalist ideology" through the Third World. In linguistic terms, by spreading standard English and French (and, to a lesser degree, German, Spanish, and Arabic), news agencies also intensified the replicative power of the norms that make up those languages. Today, for example, the linguistic flow from the Associated Press is about seventeen million words a day, most in English but some in Spanish. Reuters emits six million words a day, the majority in English but some in French and Arabic, while Agence France Presse (the successor of Havas) puts out about three million French words a day.[129]

On the other hand, large circulation newspapers (as well as advertising agencies and to a lesser extent the "telegraphic style" of the news agencies) also injected heterogeneity into the standard languages. This is only an apparent paradox, since the standards that the popular press tend to "subvert" have always been upper-class dialects, and, in their search for widened appeal, newspapers tend to use words and syntax that are not necessarily accepted as correct by that class. "Large-circulation journalism provided the means not only of renewing the language but also of sanctioning its colloquial usage and of elevating the spoken standard to the written. Journalists . . . keep close to the accents of the human voice and an oral tradition constantly informs their writing."[130] The dynamics of this heterogenization revolve around the fact that even the standard language has different registers (the formal, the colloquial, the technical), and when they meet "internal contact situations" arise. The colloquial register of the standard, for instance, is in close contact with nonstandard segments of language, such as slangs and jargons. Due to these "contact surfaces," linguistic materials elaborated as slang can flow upward through the informal register into the formal. One linguist predicts, for example, that as a result of the mass media "slang will rapidly rise to the level of the colloquial and the colloquial to the level of the standard. As a consequence of the speed-up of acceptability . . . a modern cavalier attitude towards new word formations, syntactical idioms, and specialist jargons will also intensify."[131]

Here we should bring the separate lines of our argument together. Colonialism, on one hand, and technology, on the other, greatly intensified the replicative power of the standard norms. Many regions that had formerly housed their own complex mixtures of linguistic materials were now homogenized to a certain extent by the invasion of powerful standard replicators. For the same reason, however, the standard replicators came into contact with others that, despite their low prestige, were capable of injecting them with a degree of heterogeneity. Whether the contact situations were external or internal, the effect was the same: a portion of the frozen standard was set into variation again. Further kinds of contact would soon appear as nineteenth-century technology began to affect the social structure of Europe. In particular, the growth of industrial conurbations in England (and elsewhere) and the migratory movements from the rural areas that provided coal-driven towns with workers created novel mixtures of dialects as well as a new social stratum: the industrial proletariat. The limited social mobility of these workers and their need to develop a local identity inevitably affected their transmission of linguistic norms, creating new varieties of spoken English.

In the last decades of the nineteenth century, these industrial masses came to be seen as a dangerous class, the barbarians at the gate, "creatures with strange antics and manners [who] drifted through the streets hoarsely cheering, breaking into fatuous irritating laughter, singing quaint militant songs."[132] The language of these "barbarians" was perceived by standard speakers as a nonlanguage, noisy and disarticulated, with a superabundance of negatives and a simplified grammar and vocabulary. (In short, the same traits that could be used to identify any creole around the world.) Yet, these same masses would come to be perceived as potential allies (and would eventually be granted the right to vote) when World War I transformed the new conurbations (as well as the older urban centers) into reservoirs of recruits to be conscripted.

Two education acts (one in 1870, the other in 1918) made schooling in the standard obligatory (and were correctly perceived by defenders of local dialects as an exterminating force, along with the press, railroads, tourism, and later radio). Both acts were institutional responses to the need to assimilate the masses into society, to make them "articulate," so they might better participate in democratic institutions and understand the language of command in the armed forces. The disciplinary measures envisaged by reformers included systematic training in standard sounds (leading to uniform pronunciation), lexical training (to secure clarity and correctness), and training in reading aloud (to secure proper intonation). Slang and jargon were viewed as dangerous, a "means of concealing secrets or as intentionally undignified substitutes."[133] However, the effect of compulsory education was not to erase linguistic class differences: rather than learning the "classless" standard as their exclusive new language, students of working-class background simply learned to switch codes; that is, they learned to deploy the standard in certain situations, while switching back to their native variety in their own homes and neighborhoods.[134]

Thus, universal schooling, colonialism, and early mass media, while extending the reach of the standard, also brought it into contact with other languages, codes, or registers, ensuring that it would be reinjected with heterogeneous elements and set into variation again. Given that nonstandard speakers show a greater creativity in the coining of new words and syntactical constructions, the contact between standard and non-standard speakers prevented standard languages from becoming "dead tongues," like written Latin, and connected them to fresh reservoirs of linguistic resources. However, the mere fact that a variety of linguistic replicators existed did not mean that the existing selection pressures would allow these novel variants to reenter the standard. In particular, stigmatization by speakers of the prestigious standard (and by the institutions they controlled) often kept even badly needed repairs from being selected in:

> Social influences on grammatical form may lead to situations similar to those arising from taboo in lexis... [with the difference that] the forms are rejected only in the standard language, and less in dialects. Since the standard language is thus automatically cut off from its normal sources of replenishment, its grammatical system may be left incomplete. The best-known example is the pronoun of the second person: the familiar and less polite form *thou* was replaced by the originally plural *you*, and the grammatical system has, ever since, lacked the means of distinguishing singular and plural in the second person. The reason for this is not the lack of slot-fillers, since new forms like *youse, youse 'uns, you all, y'all* have arisen to complete the system in dialect. But these forms are rejected as vulgar, and in polite English the lack has therefore to be remedied by various lexical means according to context and register, e.g. *you people, my friends, you chaps, those present.*[135]

Despite these shortcomings, it is obvious that the standardization of a language does offer "economies of scale." One economist argues, for example, that in an institutional setting bilingualism and its need for translation can be highly inefficient, involving duplication of personnel and printed material. This is particularly true of countries with a complex division of labor (with its multiplication of technical registers) and a high degree of industrialization.[136] Standardization allows a more efficient accumulation of technical vocabulary and a faster dissemination of new lexical items across the economy. Politically, a standard language also offers an efficient medium for the unification of a country and the tapping of its human resources. As the sociologist of language Joshua Fishman puts it, a standard language offers nation builders the promise of *rapid integrative returns on a large scale.*[137] It is because of these economies of scale that linguistic standardization became a central issue among nations late in achieving political unification, whether in the nineteenth century (e.g., Italy and Germany) or in the twentieth, when the colonial world broke down and the search for national unification became international.

Fishman distinguishes several roads to nationhood. On one hand, there is the road that France, England, and Spain followed, which he calls the "State-to-Nation" strategy.[138] This is the strategy followed by territories where a number of centralized (and centralizing) institutions happened to accumulate over the centuries (a royal house, centralized government traditions, educational systems, certain commercial and industrial patterns, a strong urban capital to synthesize centuries of shared experiences into a "grand tradition"). These are the countries that called themselves "historic nations," a claim to legitimacy used to justify the digestion of their minorities: Welsh, Scots, and Irish in England; Bretons, Normans, Gascons, and Occitans in France; Galicians, Catalans, and Basques in Spain. On the other hand, there are those territories that accumulated institutions, but in a decentralized pattern (Italy and Germany, and also Greece, Hungary, and Poland). These countries followed what Fishman calls the "Nation-to-State" strategy. Here, rather than a shared institutional past, ethnic uniqueness and coherence was emphasized as a form of legitimation. The people of these territories already thought of themselves as a nation (ethnically) in the process of building centralized institutions.

While those who followed the first road tended to emphasize logic and rationality as their criteria for linguistic standardization, those who followed the second route spoke of "actual usage" and "authenticity" as the only legitimate measuring rod for a national linguistic standard.[139] With the coming of the twentieth century nation building ceased to be a Western phenomenon and became the goal of every colony that had achieved its independence, of territorial entities that had never been colonized (e.g., Turkey after World War I), and even of those minorities within a state whom centralization had not managed to suppress (Irish, Bretons). In all cases, the "question of language" played a crucial role, and local languages (Turkish), lingua francas (Swahili, Malay), and even pidgins (New Guinea Pidgin, now known as neo-Melanesian) became targets for linguistic engineering and standardization.

According to Fishman, which mixture of strategies prevailed depended on whether the new countries had a single unifying tradition to use for the legitimation of their elite's projects or whether they had several or no traditions to rely on. Those who could appeal to a single grand tradition (Turkey, Israel, Thailand, Somalia, Ethiopia)[140] emphasized authenticity; those with no tradition (the Philippines, Indonesia, Tanzania, Cameroon),[141] rationality and instrumentality; while those with several competing traditions (India, Malaysia), some compromise between the two.[142] In all these cases, the process of standardization (first, "codification," or the mini-mization of variation, then "elaboration," the diversification of the institu-tional uses of the standard), which had taken centuries to achieve in England and France, was compressed into a few decades.[143]

Regardless of their different situations, these countries faced a similar challenge as they engaged in nation building: how to transform their pop-ulations into a reservoir that could be tapped for political, military, and economic mobilization. In the process of integrating their masses into a unified nation, they needed the "economies of scale" offered by standard languages. They also needed to catch up with the West as far as enrich-ing their vocabularies to confront the complexities of new technologies and organizational strategies (especially in the military, but also in corpo-rations), and this they could do either by borrowing words (as English did centuries earlier, when it was a minor language) or by developing the indigenous word-forming resources of their own standards.

While the old colonies were trying to achieve the same efficiencies of standardization as their ex-colonial masters, the languages of the two "linguistic superpowers" (French and English) were competing to become the first global superstandard. Before World War II, French was without question the international standard, having already become the language of many elites around the world and hence the most prestigious medium for diplomatic and cultural communication. Although certain setbacks in the late nineteenth century had diminished French prestige (such as the defeat to Prussia in 1870–1871), France had again emerged after World War I as the cultural center of the world. Because of its long-standing lin-guistic preeminence, France had not felt the need to create special insti-tutions to disseminate its standard around the world, with the possible exception of the Alliance Française, which was established in the 1890s. Yet, after their armies were shattered by the Nazis and their country was isolated from the outside world for several years, French speakers emerged in 1945 to confront a different linguistic situation: English was now *the* language of science and technology, and it was beginning to chal-lenge French as the chosen language of the world's elites. (Russian, too, began to replace French among the Eastern European elites who had been pulled into the Soviet sphere of influence.)[144]

France's loss of its former colonies (Lebanon and Syria by 1946, Indo-china by 1954, Tunisia and Morocco by 1956, Algeria by 1962) was an added blow to the global prestige of its language, although English was also suffering similar setbacks around the world. Americanisms, which had begun to infiltrate British English after World War I, were now invad-ing France at what seemed to the French an alarming rate. "Areas of greatest infection were sports, the world of beauty parlors (magazines such as *Elle*), toy stores and dancing."[145] French grammar itself was being penetrated: *k* and *y* entered some spellings, the form of the plural became somewhat inconsistent, and affixes such as "-rama," "super-," and "auto-" enjoyed great diffusion among the French population of replicating norms. By the early 1950s, over 20 percent of all books were published in English (less than 10 percent in French), and 50 percent of the world's newspapers and 60 percent of the world's broadcasts were in English.[146]

In response to these circumstances, when Charles de Gaulle returned to power, "France began to embark upon a positive and aggressive policy in regard to the radiation of French."[147] In 1966 a public organization was formed specifically to promote the diffusion of French (Haut Comité pour la Défense de la Langue Française), a year after Lyndon Johnson inaugurated an official campaign to teach American English abroad. Documents from these years articulate the official stance toward linguistic radiation in the same terms in which the French language had been viewed since Louis XIV: a language embodying "eternal values" (such as clarity and lack of ambiguity) and "universality" (referring to a human condition beyond time and space). Hence, imposing French on other peoples was not a form of linguistic imperialism but part of the civilizing mission of France, a liberation of those peoples from their backward provincialism.[148] Of course, given that French is a hybrid (of Mediterranean and Germanic materials) and that the Parisian dialect won its place through power, this legitimizing narrative was a fabrication by the elites. Nevertheless, the policy paid off: in 1967, thanks to the votes of France's former African colonies, French was accepted on the same level as English in the United Nations. (In 1945, to the great embarrassment and shock of French speakers, their language had been acknowledged by the U.N. as *one among many*, by a margin of only one vote.)[149]

We have already discussed the different colonialist attitudes toward local languages, and noted that the French generally assumed a more aggressive stance than the British or Germans. Robert Phillipson's analysis of linguistic imperialism accepts this to be true in the case of Africa but warns against oversimplifying the question. (For example, if one compares French Indochina to British India, the roles seemed to be reversed, with the French displaying more tolerance of indigenous languages than the British.)[150] Phillipson also argues that, even though the two linguistic superpowers have ceased to dominate their former colonies politically, they still have homogenizing effects on their cultures through the educational systems both superpowers are spreading throughout the developing nations with funds from their governments. "Just as schools were the principal instrument for alienating indigenous minorities from their languages and traditional cultures (as in the case of the Welsh, the American native peoples, and the Australian Aborigines), it is schools in Africa which are stifling local languages and imposing alien tongues and values."[151] Although Phillipson admits that, unlike French, no "master plan" for the spread of English was ever articulated in British or American institutions, the growth of English teaching as a profession, "monolingual and anglocentric, and [tending] to ignore the wider context of its operations," produced homogenizing effects in which English tended to replace or displace other languages.[152]

In addition to the educational push, big business fostered the spread of English and French, bolstering their status as international standards. I have already mentioned the international news agencies, the "big four" wholesalers of linguistic materials: Reuters, AP, UPI, and the French AFP. (These corporations also manage large flows of images, but textual news continues to be their core business.) To grasp the intensity of the linguistic flow they handle one need only learn that a subscriber to all four news services would receive on average 300,000 words a day. And technology is further intensifying this flow: while the old Teletype delivered 60 words per minute, today's computers and satellites allow 1,200 words per minute to cross continents in a format that can be fed directly into a newspaper's computerized typesetter.[153]

Since the nineteenth century, news agencies have divided the world among themselves: at present, francophone Africa belongs to AFP; anglophone Africa to Reuters; Latin America to AP and UPI. Elsewhere they engage in fierce rivalry, but of course this is oligopolistic competition, not real market competition. The news agencies have come to embody a true antimarket structure, that is, one dominated by managerial hierarchies and not by owners or their representatives.[154] Although they are not engaged in a conspiracy to promote "capitalist values" around the world, they do have a strong homogenizing effect, arising from the routinization and standardization of point of view (with the concomitant distorting simplification) and, ultimately, from the very *form of the flow*, that is, a flow emanating from very few places to a large number of subscribers. This type of flow (a "one-to-many" flow) guarantees that there will be a small number of producers of this type of "linguistic product" and a large number of consumers. The one-to-many structure of news delivery was eventually built directly into the technological infrastructure used to manage the flow. In the 1950s, for example, Reuters' financial services division began to build its own (Teletype-based) communications network for the delivery of its product (commodity and stock market news). By 1963, the

International Financial Printer began operations, but the real takeoff did not occur until the slow, bulky, and noisy teleprinters were replaced by video terminals in the 1970s. (By 1982, Reuters alone had over thirty thousand terminals in eighty-one countries.)[155]

However, by the time this one-to-many network matured, other networks began offering the possibility of a radically different paradigm: the many-to-many delivery system made possible by the Internet, the largely self-organized international meshwork of computers which formed over the past two decades. Although the Internet (or rather its precursor, the Arpanet) was of military origin (and its decentralized design a way to make it resistant to nuclear attack), the growth of its many-to-many structure was not something commanded into existence from above but an appropriation of an idea whose momentum sprang from a decentralized, largely grassroots movement. Howard Rheingold, in his history of the Internet, has brought to light the way in which geographically dispersed communities emerged as computerized communications, originally intended for technical (scientific or military) communication, were transformed into a medium supporting a variety of different forms of *conversation*. One example is the so-called Usenet, a discussion system originally designed for technical support but quickly adapted by its users for many other purposes:

> Usenet is a place for conversation or publication, like a giant coffeehouse with a thousand rooms; it is also a worldwide digital version of the Speaker's Corner in London's Hyde Park, an unedited collection of letters to the editor, a floating flea market, a huge vanity publisher, and a coalition of every odd special-interest group in the world. It is a mass medium because any piece of information put onto the Net has a potential worldwide reach of millions. But it differs from conventional mass media in several respects. Every individual who has the ability to read a Usenet posting has the ability to reply or to create a new posting. In television, newspapers, magazines, films, and radio, a small number of people have the power to determine which information should be made available to the mass audience. In Usenet, every member of the audience is also potentially a publisher. Students at universities in Taiwan who had Usenet access and telephone links to relatives in China became a network of correspondents during the 1989 Tiananmen Square incident.... Usenet is an enormous volunteer effort. The people who created it did so voluntarily and put the software in the public domain. The growing megabytes of content are contributed by volunteers.[156]

The main effect of the Internet's many-to-many structure, in terms of the fate of linguistic replicators, may be its potential for a demassification of the population, that is, its potential to create small, geographically diverse communities linked by common interests and informal conversations. Had the traffic in computer networks been dominated by the exchange of military or scientific information, we would expect to see a much higher degree of formality in the norms circulating through computers. But because the network was transformed into a conversational medium by its own users (not only English speakers but French speakers too, who transformed a one-to-many data delivery service, Minitel, into a many-to-many chat system[157]), we may speculate that the colloquial register will be strengthened by the new medium, and this despite the fact that the Internet transports mostly written text. (For instance, on one real-time chat system, the IRC, correcting misspellings as one writes is considered bad form; hence the enforcement of standard spelling, and even grammar, is weak or nonexistent.)

While the vast amounts of linguistic replicators that circulate through the Internet are therefore bound to be colloquial English, they are nevertheless *English*, which raises a number of questions. On one hand, there is nothing surprising about this since English long ago (since at least World War II) became the international lingua franca of high technology. As one author puts it, "When a Russian pilot seeks to land at an air field in Athens, Cairo or New Delhi, he talks to the control tower in English."[158] Similarly, for reasons having very little to do with its linguistic properties, English became the language of computers, both in the sense that formal computer languages that use standard words as mnemonic devices (such as Pascal or Fortran) use English as a source and in the sense that technical discussions about computers tend to be conducted in English (again, not surprisingly, since Britain and the United States played key roles in the development of the technology). On the other hand, countering the linguistic homogenization that this implies, due to its role as a lingua franca, English is being changed and adapted by foreign users in many different ways, particularly when it is taken as a source of loan words. The Japanese are famous for the way they miniaturize what they borrow from English: "modern girl" becomes "moga," "word processor" is shortened to "wa-pro," and "mass communications" to "masu-komi."[159]

The international communities that today flourish on the Internet may one day create another English, one where Japanese miniaturizations are welcomed (and so everyone engages in *masu-koming* instead of mass-communicating), where pride of the standard is seen as a foreign emotion, where a continuum of neo-Englishes flourishes, protected from the hierarchical weight of "received pronunciations" and official criteria of correctness. This would, of course, depend on how many other countries embrace the Internet as a means to build nonnational (and nonnationalistic) communities.[160] But it will also depend on what *kind of internationalism* becomes predominant on the Internet itself. As we observed in the first chapter, as antimarket institutions became international they launched an attack on national governments. The central state, a cherished partner of antimarkets for so long, suddenly became a rival and an obstacle to international expansion. Although antimarket institutions had an early presence in the computer meshwork, today they are set to invade the Internet with unprecedented force.[161] It is possible that the meshworks that have already accumulated within the Internet will prove resilient enough to survive the attack and continue to flourish. It is also possible in the next decades that hierarchies will instead accumulate, perhaps even changing the network back into a one-to-many system of information delivery. The outcome of this struggle has certainly not been settled.

Perhaps the most important lesson to be learned from the Internet experience may be that the possibilities of demassification which it has opened up have, in a sense, very little to do with futuristic technology. Although many see this computer meshwork principally as a valuable reservoir of information, its main contribution may one day be seen as a catalyst for the formation of communities (and hence as a reservoir of emotional, technical, and other types of *support*). Since communities bound by common interests existed long before computers, it is not as if we have now entered the next stage in the evolution of society (the "information age"). Rather, computer meshworks have created a bridge to a stable state of social life which existed before massification and continues to coexist alongside it. The effects of one-to-many mass media made this adjacent stable state hard to reach, but they did not leave it behind as a "primitive" form of organization. Humanity has never been moving "vertically" up a ladder of progress, but simply exploring "horizontally" a space of possibilities prestructured by stable states.

No doubt, the different dynamical processes that have shaped human history are changing this space as we move, new stable states appearing while others disappear or lose stability. The stable state defining a community of mutually supporting members obviously had not disappeared, rather we had drifted away from it, and computer networks may now bridge that gap. On the other hand, if the value of computer networks is this (nonfuturistic) catalytic role, their future worth will depend entirely on the *quality* of the communities that develop within them. Moreover, these communal meshworks will embrace people with diverse political inclinations (including fascistic communities), so that the mere existence of "virtual communities" will not guarantee social change in the direction of a fairer, less oppressive society. To paraphrase Deleuze and Guattari, never believe that a meshwork will suffice to save us.[162]

Conclusion and Speculations

In terms of the nonlinear dynamics of our planet, the thin rocky crust on which we live and which we call our land and home is perhaps the earth's least important component. The crust is, indeed, a mere hardening within the greater system of underground lava flows which, organizing themselves into large "conveyor belts" (convective

cells), are the main factor in the genesis of
the most salient and apparently durable
structures of the crusty surface. Either
directly, via volcanic activity, or indirectly, by
forcing continental plates to collide, thereby
creating the great folded mountain ranges, it
is the self-organized activity of lava flows that
is at the origin of many geological forms. If
we consider that the oceanic crust on which
the continents are embedded is constantly
being created and destroyed (by solidification
and remelting) and that even continental
crust is under constant erosion so that its
materials are recycled into the ocean, the
rocks and mountains that define the most
stable and durable traits of our reality would
merely represent a local *slowing down* of this
flowing reality. It is almost as if every part
of the mineral world could be defined simply
by specifying its chemical composition and
its *speed of flow*: very slow for rocks, faster
for lava.

Similarly, our individual bodies and minds
are mere coagulations or decelerations in the
flows of biomass, genes, memes, and norms.
Here, too, we might be defined both by the
materials we are temporarily binding or chain-

ing to our organic bodies and cultural minds and by the timescale of the binding operation. Over the millennia, it is the flow of biomass through food webs, as well as the flow of genes through generations, that matters, not the bodies and species that emerge from these flows. Our languages may also be seen over time as momentary slowing downs or thickenings in a flow of norms that gives rise to a multitude of different structures. And a similar point applies to our institutions, which may also be considered transitory hardenings in the flows of money, routines, and prestige, and, if they have acquired a permanent building to house them, in the mineral flows from which the construction materials derive.

This book has concerned itself with a historical survey of these flows of "stuff," as well as with the hardenings themselves, since once they emerge they react back on the flows to constrain them in a variety of ways. Although this simple statement captures the gist of the book, it must be qualified in several ways. On one hand, the flows of materials whose history we described involved more than just matter-energy. They also included *information*, understood not in static terms as mere physical patterns (measured in bits) but in dynamic terms, as patterns capable of self-replication and catalysis. That is, we have considered not only genes, memes, and norms, but also the "phenotypic" effects of these replicators, their ability to trigger intensifications or diminutions in the flows of matter-energy and their ability to switch from one stable state to another the structures that emerge out of these flows. On the other hand, among these structures we distinguished coagulations that have undergone a process of homogenization, which we called *hierarchies* (or more generally, *strata*), from those wherein heterogeneous components were articulated as such, which we referred to as *meshworks* (or more generally, *self-consistent aggregates*).

We repeatedly saw that hierarchies and meshworks occur mostly in mixtures, so it is convenient to have a label to refer to these changing combinations. If the hierarchical components of the mix dominate over the meshwork components, we may speak of a highly *stratified* structure, while the opposite combination will be referred to as having a low degree of stratification. Moreover, since meshworks give rise to hierarchies and hierarchies to meshworks, we may speak of a given mixture as undergoing processes of *destratification* as well as *restratification*, as its proportions of homogeneous and heterogeneous components change. Finally, since what truly defines the real world (according to this way of viewing things) are neither uniform strata nor variable meshworks but the unformed and unstructured flows from which these two derive, it will also be useful to have a label to refer to this special state of matter-energy information, to this flowing reality animated from within by self-organizing processes constituting a veritable *nonorganic life*: the Body without Organs (BwO). As Gilles Deleuze and Félix Guattari write:

> The organism is not at all the body, the BwO; rather it is a stratum on the BwO, in other words, a phenomenon of accumulation, coagulation, and sedimentation that, in order to extract useful labor from the BwO, imposes upon it forms, functions, bonds, dominant and hierarchized organizations, organized transcendences.... [T]he BwO is that glacial reality where the alluvions, sedimentations, coagulations, foldings, and recoilings that compose an organism—and also a signification and a subject—occur.[1]

The label itself is, of course, immaterial and insignificant. We could as well refer to this cauldron of nonorganic life by a different name. (Elsewhere, for instance, we called it the "machinic phylum.")[2] Unlike the name, however, the referent of the label is of extreme importance, since the flows of lava, biomass, genes, memes, norms, money (and other "stuff") are the source of just about every stable structure that we cherish and value (or, on the contrary, that oppresses or enslaves us). We could define the BwO in terms of these unformed, destratified flows, as long as we keep in mind that what counts as destratified at any given time and in any given space is entirely relative. The flow of genes and biomass are "unformed" if we compare them to any individual organism, but the flows themselves have internal forms and functions. Indeed, if instead of taking a planetary perspective we adopted a cosmic viewpoint, our entire planet would itself be a mere provisional hardening in the vast flows of plasma which permeate the universe.

Plasmas, clouds of electrified elementary particles that have lost even their atomic forms, are (as far as we know) the state of matter-energy with the least amount of internal structure, and yet they are capable of supporting a variety of self-organizing processes. However, rather than identifying the BwO with the plasma that fills our universe, we should think of it as *a limit of a given process of destratification*: plasmas may indeed be such a limit when we think of mineral structures, but not if we think of genetic materials. The more or less free and unformed flow of genes through microorganisms may be a better illustration of what the BwO of a flow of replicators may be. On the other hand, an egg (and all the self-organizing processes that animate its cytoplasm) is a good image of a BwO in the flow of biomass: an unformed body of energetic and mineral materials having the potential to give rise to a variety of organs once it is fertilized and begins developing into an embryo.[3]

It would, of course, be possible to frame my concluding remarks without using these terms, and throughout this book I have attempted to carry on my argument with a minimum of strange-sounding jargon. There are, however, two advantages to introducing these terms at this point. First, they allow for a more compact description: any structure that matters as far as human history is concerned may be defined by its degree of stratification, and changes in composition between command and market components may be defined as movements of destratification and restratification. Second, having established the plausibility of this philosophical stance through an analysis that never strayed far from historical realities, this compact set of terms will allow me to conclude this discussion in a more speculative vein while keeping it from drifting away from the ideas we have already explored.

Human history has involved a variety of Bodies without Organs. First, the sun, that giant sphere of plasma whose intense flow of energy drives most processes of self-organization on our planet and, in the form of grain and fossil fuel, of our civilizations. Second, the lava "conveyor belts" that drive plate tectonics and are responsible for the most general geo-political features of our planet, such as the breakdown of Pangaea into our current continents, and the subsequent distribution of domesticable species, a distribution that benefitted Eurasia over America, Africa, and Oceania.[4] Third, the BwO constituted by the coupled dynamics of the hydrosphere and atmosphere and their wild variety of self-organized enti-ties: hurricanes, tsunamis, pressure blocks, cyclones, and wind circuits. As we saw, the conquest of the wind circuits of the Atlantic (the trade winds and the westerlies) allowed the transformation of the American continent into a vast supply zone to fuel the growth of the European urban economy. Fourth, the genetic BwO constituted by the more or less free flow of genes through microorganisms (via plasmids and other vec-tors), which, unlike the more stratified genetic flow in animals and plants, has avoided human control even after the development of antibiotics. Fifth, those portions of the flow of solar energy through ecosystems (flesh circulating in natural food webs) which have escaped urbanization, particularly animal and vegetable weeds, or rhizomes (the BwO formed by an underground rodent city, for example).[5] Finally, our languages also formed a BwO when they formed dialect continua and circumstances conspired to remove any stratifying pressure, as when the Norman invaders of England imposed French as the language of the elites, allow-ing the peasant masses to create the English language out of an amor-phous soup of Germanic norms with Scandinavian and Latin spices. (Because all five of these BwOs, unlike pure plasmas, retain forms and functions, they may be considered examples of a local BwO, that is, local limits of a process of destratification, and not *the BwO*, taken as an absolute limit. However, for simplicity, I will continue to refer to these limit states in the singular.)

I have attempted here to describe Western history in the last one thou-sand years as a series of processes occurring in the BwO: pidginizations, creolizations, and standardizations in the flow of norms; isolations, con-tacts, and institutionalizations in the flow of memes; domestications, fer-alizations, and hybridizations in the flow of genes; and intensifications, accelerations, and decelerations in the flows of energy and materials. Cities and their mineral exoskeletons, their shortened food chains, and their dominant dialects are among the structures we saw emerge from these nonlinear flows. Once in place, they reacted back on the flows, either to inhibit them or to further stimulate them. In other words, cities appeared not only as structures operating at a certain degree of stratifi-cation (with a certain mix of market and command components), but they themselves performed destratifications and restratifications on the flows that traversed them. And a similar point applies to the populations of institutions that inhabited these urban centers as well as to their populations of human minds and bodies.

The concept of the BwO was created in an effort to conceive the genesis of form (in geological, biological, and cultural structures) as related exclusively to *immanent* capabilities of the flows of matter-energy information and not to any *transcendent* factor, whether platonic or divine. To explain this inherent morphogenetic potential without sneaking transcendental essences through the back door, Deleuze and Guattari developed their theory of abstract machines, engineering diagrams defining the structure-generating processes that give rise to more or less permanent forms but are not unique to those forms; that is, they do not represent (as an essence does) that which defines the identity of those forms. Attractors are the simplest type of abstract machine, operating at the level of nonlinear, destratified flows. Attractors represent patterns of stability and becoming that are inherent in abstract dynamical systems and may be "incarnated" in a variety of actual physical systems. For example, one and the same periodic attractor may be instantiated by lava or wind in a convection cell, the spontaneous rhythmical behavior of crystal radios, periodic behavior in electronic circuits or chemical reactions, and even the behavior of an economic system during a business cycle. A dynamical system whose behavior is governed by these endogenously generated stable states is further characterized by a certain number of key parameters. At any one moment in the system's history it is the *degree of intensity* of these parameters (the degree of temperature, pressure, volume, speed, density, and so on) that defines the attractors available to the system and, hence, the type of forms it may give rise to. (That is, at critical values of these parameters, bifurcations occur which abruptly change one set of attractors into another.)

Similar considerations apply to the more complex abstract machines that emerge from these simple ones. The two most general abstract diagrams that we examined were those behind the formation of strata and self-consistent aggregates. The hierarchy-generating machine involved a process of double articulation, that is, a sorting operation that yields a homogeneous distribution of elements and a consolidation operation that defines more or less permanent structural linkages between sorted materials. The meshwork-generating machine, on the other hand, articulates divergent but partially overlapping components by their functional complementarities, using a variety of local intercalary elements as well as endogenously generated stable states. Then we discovered that, if and when the materials on which a sorting device operates acquire the ability to replicate with variation, a new abstract machine emerges, in the form of a blind probe head capable of exploring a space of possible forms. Finally, with the creation of social networks capable of acting as enforcement mechanisms for the transmission of norms, an abstract machine operating by means of combinatorial constraints was made possible, defining the structure-generating process behind some linguistic structures.

These other abstract machines may also be viewed as equipped with "knobs," controlling parameters whose intensity defines the dynamical state of the structure-generating process. For instance, we saw that in Mary Douglas's theory of the genesis of discursive form (coherent worldviews) the intensity of allegiance to a group, as well as the intensity of outside regulation to which the group must conform (that is, the values of the group-and-grid parameters), defines the stable states to which a collectivity of believers (and their beliefs) will be drawn. In Zellig Harris's theory of language, on the other hand, the degree of variability of the operator-argument constraint, as well as the strength of constraints on sequences of inscriptions, determines whether the sequences generated will be of the logico-mathematical, linguistic, or musical type. Other key parameters are those controlling the strength and thoroughness of the sorting process and the degree of consolidation or reproductive isolation of the double-articulation machine; or the degree of connectivity that determines when a meshwork becomes self-sustaining; or the rates of mutation and recombination that define the speed of the probe head, as well as the strength of the flow of biomass and of the coupling between coevolving species—parameters that define the kind of space that the probe head explores. Hence, using these abstract diagrams to represent what goes on in the BwO is equivalent to using a system of representation in terms of *intensities*, since it is ultimately the intensity of each parameter that determines the kind of dynamic involved and, hence, the character of the structures that are generated. Indeed, one way of picturing the BwO is as the "glacial" state of matter-energy information resulting from turning all these knobs to *zero*, that is, to the absolute minimum value of intensity, *bringing any production of structured form to a halt.* As Deleuze and Guattari write:

> A BwO is made in such a way that it can be occupied, populated only by intensities. Only intensities pass and circulate. Still, the BwO is not a scene, a place, or even a support upon which something comes to pass.... It is not space, nor is it in space; it is matter that occupies space to a given degree—to the degree corresponding to the intensities produced.

It is nonstratified, unformed, intense matter, the matrix of intensity, intensity = 0.... Production of the real as an intensive magnitude starting at zero.[6]

To view human history as unfolding immersed in this cauldron of nonorganic life is one way to eliminate notions of progress or unilineal development. Indeed, the three narratives I used to approach the geological, biological, and linguistic histories of the West were framed not in terms of "man" and his manifest destiny, but in terms of stuff undergoing different kinds of *intensification*. In those three narratives we used the year 1000 as a degree zero of intensity for the West, while the powerful agricultural intensification that occurred before the turn of the first millennium was seen as having jump-started the abstract machines and begun the process of structure formation again. This intensification in turn acted as a trigger for a whole series of further intensifications: of density of settlement and degree of mineralization; of the velocity and quantity of money in circulation; of the accumulation of know-how and formal knowledge; of the formation of urban hierarchies and the proliferation of links among maritime gateways; of the divergence of spoken Latin varieties and the standardization of writing and spelling systems. Later on, intensified colonialism and conquest, routinization and rationalization, money and knowledge accumulation, and fossil-energy flow resulted in the self-sustained intensification known as the Industrial Revolution. In both periods, there were catalysts of different kinds (money, technology, know-how) effecting and sustaining the intensifications. And in both periods, the very flows triggered by one catalyst became triggers for yet other flows, the whole assemblage of trigger flows acquiring autocatalytic dynamics. (Only in the second period, it was argued, did these trigger flows form a closed circuit of enough complexity to become a self-sustaining autocatalytic loop.)

Although both meshworks and hierarchies arose from the first urban intensification (1000–1300 A.D.), the overall effect of the acceleration of city building in Europe was destratifying. As Braudel observed, urban centers in the West were veritable accelerators of historical time as well as machines for the breaking of old bonds (such as those chaining peasants to their feudal lords). But here we must be careful in our evaluations, since at all points there were coexisting histories moving at different speeds or with different degrees of destratification: the rural masses moved at one speed, the urban markets at a faster rhythm, while commercial and financial antimarkets achieved the greatest degree of mobility. For example, the flows of money that markets used to mobilize the food surpluses produced in rural regions acquired new properties in the hands of antimarket institutions, detaching themselves from any particular flow of matter-energy and invading any economic activity where profits were particularly intense.

Moreover, before the Industrial Revolution, the cutting edge of antimarket development was represented by the maritime gateways (Venice, Genoa, Amsterdam) that were the most destratified—that is, the least attached to the land for its agricultural resources (they were all ecologically deprived)—as well as the least concerned with the government and control of large territories. As Paul M. Hohenberg and Lynn Hollen Lees note, these gateways had closer ties to overseas colonies and to one another than to the territories at their backs. On the other hand, when the destratified financial flows that these gateways (and the regional capitals closely associated with them) generated were invested in mines or protoindustrial activities, the structures generated were extremely stratified and hierarchical, rivaling those of contemporaneous military institutions (such as the arsenal of Venice) in their degree of central control and industrial discipline. Deleuze and Guattari, noticing this apparent paradox, write that it was "precisely because the bourgeoisie was a cutting edge of deterritorialization, a veritable particle accelerator, that it also performed an overall reterritorialization."[7] Although their explanation of this paradox is rather complex, we can summarize it in a general hypothesis: that the creation of *novel* hierarchical structures through restratification is performed by the most destratified element of the *previous* phase.

We may agree with this important hypothesis as long as we do not view the restratified result of the powerful destratification that mobile antimarkets represented as a societywide system (capitalism), but simply as a new breed of organizations (and institutional norms) that added themselves to the existing ecology of institutions, interacting with them and the trigger flows under their control. Resisting the temptation to reduce complex institutional dynamics to a single factor (e.g., antimarket economics) is even more important when considering the great circuit of trigger flows that formed the basis for the Industrial Revolution. No doubt, antimarkets played a key role in the conjunction of trigger flows (coal, steam, cotton, iron, raw labor, skills) that made up the factory towns and the industrial conurbations. But here, too, other destratified elements, other particle accelerators were necessary: the British government destratifying its territory by abolishing tolls and tariffs and creating a national market, and destratifying its taxation and fiscal system by creating the Bank of England and the very notion of national debt. In France, the army was becoming the most destratified in Europe, leading not only to Napoleon's fantastically mobile war machine but simultaneously to a greater restratification: the conversion of warfare from the limited dynastic duels of the eighteenth century to the "total war" with which we are familiar today, involving the complete mobilization of a country's resources by a centralized governmental agency.

Moreover, not only were there several particle accelerators mobilizing trigger flows of different kinds, there were coexisting motions of destratification of *intermediate intensity* which connected these flows, generating meshworks of different kinds: peasant and small-town markets; symbiotic nets of small producers engaged in volatile trade and import substitution; large cities and industrial hinterlands operating via economies of agglomeration; alpine regions elaborating industrial paradigms different from those of the coal conurbations, in which skills and crafts were meshed together instead of being replaced by routines and centralized machinery. What use is there in moving our level of description to the BwO if we are not going to take advantage of the heterogeneous mixtures of energy and genes, germs and words, which it allows us to conceive, a world in which geology, biology, and linguistics are not seen as three separate spheres, each more advanced or progressive than the previous one, but as three perfectly coexisting and interacting flows of energetic, replicative, and catalytic materials? What use is there in making this move, if we are to crown the whole exercise with a return to the great master concept, the great homogenization involved in the notion of a "capitalist system"? On the contrary, we must be cautious when deploying our concepts, not only when we periodize human history, but also when we think of our evolution from geologic and organic strata:

> It is difficult to elucidate the system of the strata without seeming to introduce a kind of cosmic or even spiritual evolution from one to the other, as if they were arranged in stages and ascending degrees of perfection. Nothing of the sort.... If one begins by considering the strata in themselves, it cannot be said that one is less organized than the other.... [T]here is no fixed order, and one stratum can serve directly as a substratum for another without the intermediaries one would expect from the point of view of stages and degrees.... Or the apparent order may be reversed, with cultural or technical phenomena providing a fertile soil, a good soup, for the development of insects, bacteria, germs or even particles. The industrial age defined as the age of insects.... [On the other hand] if we consider the plane of consistency [the BwO at the absolute limit of destratification] we note that the most disparate things and signs move upon it: a semiotic fragment rubs shoulders with a chemical interaction, an electron

clashes into a language.... There is no "like" here, we are not saying "like an electron," "like an interaction," etc. The plane of consistency is the abolition of all metaphor; all that consists is Real.[8]

Thus, according to Deleuze and Guattari, in terms of the stratified and the destratified, human history is not marked by stages of progress but by coexistences of accumulated materials of diverse kinds, as well as by the processes of stratification and destratification that these interacting accumulations undergo. In this sense, we could characterize our era as the "age of information" or, equally validly, as the "second age of insects and germs," given the significant accumulations of insecticide- and antibiotic-resistant genes which our agricultural and medical practices have inadvertently fostered since World War II. And as I attempted to show in this book, these mixtures of coexisting "ages" are not something new but have always characterized human history. Medieval towns were both linguistic and epidemiological laboratories, and many things accumulated within their walls: money, skills, weeds, cattle, manuscripts, prestige, power. In the nineteenth century, as cholera epidemics were giving rise to public health organizations, the inanimate power of coal and steam was transforming the world not only into a single disease pool, but also into a single norm pool (at least for some major languages, such as English and French), and, of course, a single world-economy. Rats and their fleas and germs were traveling in the same transoceanic ships that brought to the neo-Europes millions of people, as well as a great variety of other stuff: raw materials, silver, luxury items, domesticated species, investment capital, weapons, and so on.

In turn, the nineteenth century witnessed the proliferation of institutions dedicated to disentangling these dangerous mixtures: naval hospitals and schools, prisons, and factories. These institutional sorting devices began to process particular flows and to assign each geological, biological, and linguistic component its "proper" place. As Foucault has shown, the sorting operation was carried out in these institutions via spatial partitioning and standardized tests of different kinds, as well as by an elaborate record-keeping system to store the results of those examinations. In terms of abstract diagrams, there is no difference between these institutions and the rivers that sort out the sediment that forms certain rocks, or the ecological selection pressures that sort genes into species. (That is, in all three cases we have an operation of sorting, classifying, or "territorializing.") But what would correspond to the second operation, the cementing together of the sediment or the reproductive isolation of the species acting as a ratchet mechanism? (That is, the "coding" of permanent architectural relations between pebbles, or the "coding" of a species as a separate reproductive entity through changes in its mating calls, scent, or visual markings.)

In his reading of Foucault, Deleuze has given us some clues regarding this second articulation. He uses the terms "content" and "expression" to refer to the two articulations and warns us not to confuse them with the old philosophical distinction between substance and form. Instead, each articulation includes both forms and substances: sediment is not only an accumulation of pebbles (substance), it is an accumulation distributed in homogeneous layers (form); in turn, cementing these pebbles together establishes spatial links among pebbles (form) and creates a material entity of a larger scale, a sedimentary rock (substance). The same holds true for institutional entities, such as hospitals, schools, and prisons:

> Strata are historical formations.... As "sedimentary beds" they are made out from things and words, from seeing and speaking, from the visible and the sayable, from bands of visibility and fields of readability, from contents and expressions.... The content has both a form and a substance: for example, the form is the prison and the substance is those who are locked up, the prisoners.... The expression also has a form and a substance: for example, the form is penal law and the substance is "delinquency" in so far as it is the object of statements.[9]

Although the sorting operations carried out in hospitals, schools, barracks, and prisons involved different types of examination (not just visual examinations), the homogenizations they effected on the flow of human bodies were indeed intended not to fuse those bodies into an undifferentiated mass but, on the contrary, *to make visible* their individual differences so that they could be properly distributed into the ranks of the new meritocracies. At the same time—in a *distinct and separate* operation, particular discourses (medical, pedagogical, penal) were generated in and around these institutional setups and codified and consolidated the results of the sorting process into larger-scale entities: organized medicine and the educational and penal systems. (These entities were isomorphic with sedimentary rock, using "visibilities" as their pebbles and "sayabilities" as their cement.)

But it would be wrong to think that strata are the last word in this respect. Even if we agree that certain institutions played the role of first articulation, and that certain types of knowledge performed the second one, this would only give us an account of one form of power and knowledge, *formal* power and knowledge. But in addition to stratified, formal power, there is power of the meshwork type, that is, destratified power operating via a multiplicity of informal constraints. In this book we treated these constraints as catalysts, or triggers, that play the role of intercalary elements in the formation of meshworks. Although in chemistry the function of catalysts is viewed simply in terms of inhibition or stimulation, in the more abstract sense in which I have used the term the number of different constraining functions that a trigger may play should be conceived "as a necessarily open list of variables expressing a relation between forces... constituting actions upon actions: to incite, to induce, to seduce, to make easy or difficult, to enlarge or limit, to make more or less probable, and so on."[10]

Other abstract machines must be added to those behind the genesis of meshworks and hierarchies to give a fuller characterization of the complex history of Western institutional ecologies. As we noted, many hierarchical organizations with routinized activities may use these routines as a kind of "organizational memory." When these organizations reproduce (i.e., when a commercial organization opens a new branch or a government institution replicates in its overseas colonies), these routines are passed on to their progeny with variation, and this allows populations of such organizations to embody an abstract probe head. A similar point applies to institutional norms (legal, commercial, linguistic) transmitted through social obligation and to informal behavioral patterns (memes) replicated through imitation. Additionally, models like those created by Douglas suggest that yet other structure-generating processes may operate *within* formal organizations and informal networks, affecting the way in which their constituent individuals interact, and the worldviews those interactions generate.

This list of abstract machines is probably not exhaustive; there may certainly be others, governing dynamics in areas outside the scope of this book. And, indeed, even in the areas we did explore there may be alternatives (or additions) to the diagrams here proposed. But whether these or other diagrams are used to model the structure-generating processes involved in the genesis of social forms, what matters is explaining this genesis in an entirely bottom-up way. That is, not simply to assume that society forms a system, but to account for this systematicity as an emergent property of some dynamical process. This is very different from the top-down method that orthodox sociologists and other social scientists use when they begin their analysis at the level of society as a whole, justifying that approach either by using the individual organism as a metaphor for society, as in functionalist sociology, or on the basis of an imaginary dynamics, as in Marxist sociology's dialectics. On the other hand, the opposite mistake (illustrated by orthodox microeconomics) must also be avoided: atomizing society into a set of independently acting individuals. Rather, we must take into account that the larger-scale structures that emerge from the actions of individual decision makers, such as formal organizations or informal networks, have a life of their own. They are wholes that are more than the sum of their parts, but wholes that *add themselves to an existing population of individual structures,* operating at different scales (individual institutions, individual cities, individual complexes of cities, and so on). As Deleuze and Guattari put it:

> We no longer believe in a primordial totality that once existed, or in a final totality that awaits us at some future date. We no longer believe in the dull gray outlines of a dreary, colorless dialectic of evolution, aimed at forming a harmonious whole out of heterogeneous bits by rounding off their rough edges. We believe only in totalities that are peripheral. And if we discover such a totality alongside various separate parts, it is a whole *of* these particular parts but does not totalize them; it is a unity *of* all those particular parts but does not unify them; rather it is added to them as a new part fabricated separately.[11]

From the perspective of a bottom-up methodology, it is incorrect to characterize contemporary societies as "disciplinary," or as "capitalist," or, for that matter, "patriarchal" (or any other label that reduces a complex mixture of processes to a single factor), unless one can give the details of a structure-generating process that results in a societywide system. Certain institutional forms may indeed proliferate in a population, but even when this leads to the extinction of prior forms this should not be treated as the achievement of a new unified stage of development. Moreover, a given proliferation of institutions may be the result of an intensification of previously existing processes. In the case of utilitarian rationalization, as Foucault says, "the classical age did not initiate it; rather it accelerated it, changed its scale, gave it precise instruments."[12] No doubt, an intensification may lead to the crossing of a threshold, as in the critical point of complexity at which autocatalytic loops become self-sustaining, leading to industrial takeoff. Or it may lead to the creation of truly novel types of institution. But the resulting emergent structures simply add themselves to the mix of previously existing ones, interacting with them, but never leaving them behind as a prior stage of development (although, perhaps, creating the conditions for their disappearance).

This brings us to the question of the pragmatic uses of these ideas. The last three or four centuries have witnessed an intense homogenization of the world (biologically, linguistically, economically), a fact that in itself would seem to recommend the injection of a healthy dose of heterogeneity into the mix. Or, in the shorthand we have been using, the world has become so greatly stratified that the only way out is to destratify it. But there are several things wrong with this knee-jerk response.

First, although it is true that nation-states swallowed their minorities and digested them by imposing national standards for language, currency, education, and health, the solution to this is not simply to break up these large sociopolitical entities into smaller ones (say, one for each minority: for instance, the way Yugoslavia was broken up into territories for Serbs, Croats, and other minorities). To simply increase heterogeneity without articulating this diversity into a meshwork not only results in further conflict and friction, it rapidly creates a set of smaller, internally homogeneous nations. (Hence, the balkanization of the world would increase heterogeneity only in appearance.)

Second, even if we manage to create local connections between heterogeneous elements, the mere presence of an emergent meshwork does not in itself mean that we have given a segment of society a less oppressive structure. The nature of the result will depend on the character of the heterogeneous elements meshed together, as we observed of communities on the Internet: they are undoubtedly more destratified than those subjected to massification by one-to-many media, but since everyone of all political stripes—even fascists—can benefit from this destratification, the mere existence of a computer meshwork is no guarantee that a better world will develop there. Finally, increasing the proportion of meshwork in the mix is indeed destratifying, but we still need to be cautious about the speed and intensity of this destratification, particularly if it turns out to be true that "the most destratified element in a mix effects the most rigid restratification" later on:

> You don't reach the BwO, and its plane of consistency, by wildly destratifying.... If you free it with too violent an action, if you blow apart the strata without taking precautions, then instead of drawing the plane you will be killed, plunged into a black hole, or even dragged towards catastrophe. Staying stratified—organized, signified, subjected—is not the worst that can happen; the worst that can happen is if you throw the strata into demented or suicidal collapse, which brings them back down on us heavier than ever. This is how it should be done: lodge yourself on a stratum, experiment with the opportunities it offers, find an advantageous place on it, find potential movements of deterritorialization, possible lines of flight, experience them, produce flow conjunctions here and there, try

CONCLUSION AND SPECULATIONS

out continuums of intensities segment by segment, have a small plot of
new land at all times.[13]

All these precautions are necessary in a world that does not possess a
ladder of progress, or a drive toward increased perfection, or a promised
land, or even a socialist pot of gold at the end of the rainbow. Moreover,
these warnings derive from a recognition that our world is governed not
only by nonlinear dynamics, which makes detailed prediction and control
impossible, but also by nonlinear combinatorics, which implies that the
number of possible mixtures of meshwork and hierarchy, of command
and market, of centralization and decentralization, are immense and that
we simply cannot predict what the emergent properties of these myriad
combinations will be. Thus the call for a more *experimental* attitude
toward reality and for an increased awareness of the potential for self-
organization inherent in even the humblest forms of matter-energy.

When we think that the majority of equations used in science are linear and that a linear conception of causality dominated Western thought
for over two millennia, we may be inclined to think that our lack of familiarity with questions of self-organized heterogeneity and our tendency to
think about complexity in terms of homogeneous hierarchies derive from
the way *we represent the world to ourselves*. No doubt, the entrenchment
in the academic and scientific worlds of certain discursive practices
informed by linear thinking and linear representation is indeed part of
our problem. But to try to reduce a complex situation to a question of
representations is, in turn, a homogenizing force very much alive today
among social critics. Here we have argued that *both the world of objective
referents and the world of labels and concepts have undergone processes of
uniformation and standardization,* so that both discursive and nondiscursive practices need to be taken into account when tracing the history of
our homogenization.

In short, as our industrial, medical, and educational systems became
routinized, as they grew and began to profit from economies of scale, linear equations accumulated in the physical sciences and equilibrium theories flourished in the social sciences.[14] In a sense, even though the world
is inherently nonlinear and far from equilibrium, its homogenization
meant that those areas that had been made uniform began *behaving
objectively* as linear equilibrium structures, with predictable and control-
lable properties. In other words, Western societies transformed the objective world (or some areas of it) into the type of structure that would
"correspond" to their theories, so that the latter became, in a sense, self-
fulfilling prophecies.

273

Today, our theories are beginning to incorporate nonlinear elements, and we are starting to think of heterogeneity as something valuable, not as an obstacle to unification. Negative and positive feedback have been added to older linear notions of causality, enriching our conceptual reservoir. Even some materials (such as fiberglass and other composites) have increased our awareness of the limitations imposed by uniformity and our awareness of the great advantages of meshworks in industrial design.[15] In short, our theories are shedding some of their homogeneity. Although this is a welcome development, we still have to deal with the world of referents, with the thousands of routinized organizations that have accumulated over the years, with the spread of standardized languages, and with the homogenized gene pools of our domestic plants and animals, to mention only the examples discussed in this book. Changing our way of thinking about the world is a necessary first step, but it is by no means sufficient: we will need to *destratify reality itself*, and we must do so without the guarantee of a golden age ahead, knowing full well the dangers and possible restratifications we may face.

It is important, however, not to confuse the need for caution in our exploration of the nonlinear possibilities of (economic, linguistic, biological) reality, and the concomitant abandonment of utopian euphoria, with despair, resentment, or nihilism. There is, indeed, a new kind of hope implicit in these new views. After all, many of the most beautiful and inspiring things on our planet may have been created through destratification. A good example of this may be the emergence of birdsongs: the mouth became destratified when it ceased to be a strictly alimentary organ, caught up in the day-to-day eating of flesh, and began to generate other flows (memes) and structures (songs) where the meshwork element dominated the hierarchical.[16] The emergence of organic life itself, while not representing a more perfect stage of development than rocks, did involve a greater capacity to generate self-consistent aggregates, a surplus of consistency.[17] The human hand may also have involved a destratification, a complete detachment from locomotive functions and a new coupling with the external environment, itself further destratified when the hand began converting pieces of it (rocks, bones, branches) into tools.[18] Thus, despite all the cautionary tales about simplistic calls for anarchic liberation, there is in these new theories a positive, even joyful conception of reality. And while these views do indeed invoke the "death of man," it is only the death of the "man" of the old "manifest destinies," not the death of humanity and its potential for destratification.[19]

Notes

INTRODUCTION

1. See Ilya Prigogine and Isabelle Stengers, *Order Out of Chaos: Man's New Dialogue with Nature* (New York: Bantam, 1984). Prigogine and Stengers write: "We have seen new aspects of time being progressively incorporated into physics, while the ambitions of omniscience inherent in classical science were progressively rejected.... Indeed, history began by concentrating mainly on human societies, after which attention was given to the temporal dimensions of life and geology. The incorporation of time into physics thus appears as the last stage of a progressive reinsertion of history into the natural and social sciences" (p. 208).

On the role of minor fluctuations in determining the future history of a system after a bifurcation, see *ibid.*, ch. 6.

2. The term "fitness" has in fact changed in meaning with neo-Darwinism. In the nineteenth century it denoted a set of aptitudes and adaptive traits necessary for survival; today it simply means fertility or, rather, the number of offspring reared to reproductive age. This has taken away the heroic connotations of the term "fittest," which is what Social Darwinists exploited in their racist theories. It has also made it relatively straightforward (if somewhat tautological) to define optimal fit: the genes that survive are the ones that create more reproducible copies of themselves. In this sense, optimality (and the limited role for history that it involves) may still have a place in evolutionary theory. But when it comes to optimality of *adaptive traits not directly related to reproduction*, the idea that natural selection can sculpt plant and animal bodies that are optimally adapted to their environments has been losing ground. In particular, the introduction of theories of self-organization and nonequilibrium, nonlinear dynamics into the formalism of neo-Darwinism, has made it clear that selection pressures cannot achieve optimal results, particularly in cases of coevolution, as in predator-prey arms races. On the other hand, some scientists (Brian Goodwin and Francisco Varela, for example) are

so impressed by spontaneous morphogenesis that they err in the opposite direction, denying any important role to natural selection. Here I will adopt the position that both selection *and* self-organization matter in the accumulation of adaptive traits, as argued in Stuart Kauffman, *The Origins of Order: Self-Organization and Selection in Evolution* (New York: Oxford University Press, 1993), esp. ch. 3.

It is from Kauffman that I take the term "meshwork," which figures prominently throughout this book. The term appears in Stuart Kauffman, "Random Grammars: A New Class of Models for Functional Integration and Transformation in the Biological, Neural and Social Sciences," in *1990 Lectures in Complex Systems*, eds. Lynn Nadel and Daniel Stein (Redwood City, CA: Addison-Wesley, 1991), p. 428.

As important as Kauffman's work has been in giving self-organization a place in evolutionary theory, Kauffman still seems married to an old philosophy of science according to which scientists discover "universal laws" that, together with a description of initial and boundary conditions, can then be used to derive predictions purely mechanically (i.e., by means of deduction). This philosophical inheritance from the now-defunct positivist movement (called the nomological-deductive model of scientific explanation) needs to be rejected to take full advantage of the new knowledge generated by nonlinear science. On all this, and for a full explanation of what the new paradigm means for biology and philosophy of science, see David J. Depew and Bruce H. Weber, *Darwinism Evolving: Systems Dynamics and the Genealogy of Natural Selection* (Cambridge, MA: MIT Press, 1995), esp. chs. 13–18.

3. Arthur Iberall, "A Physics for the Study of Civilizations," in *Self-Organizing Systems: The Emergence of Order*, ed. Eugene Yates (New York: Plenum, 1987), pp. 531–33.

4. Arthur Iberall, *Toward a General Science of Viable Systems* (New York: McGraw-Hill, 1972), pp. 211 and 288. In this work, Iberall views the transition from agricultural to urban humanity as a bifurcation from a "liquid-droplet" phase to a "plastic-solid" one (p. 211). In his later work, this transition is viewed differently. The switch to urban life is viewed less as a result of the crystallization of a bureaucratic elite and instead as caused by the self-organization of trade flows between a small number of settlements. Or, in physical terms, trade among a small number of liquid settlements is viewed as creating a self-sustaining *convection cell*, similar to the ones that give rise to periodic winds (e.g., trade winds, monsoons). In short, while in his early work Iberall viewed the coming of urban life as an *equilibrium phase transition* (albeit involving nonequilibrium structures, such as elites playing the role of dislocations [*ibid.*, p. 208]), he later thought of it as a *nonequilibrium transition*: "But simply because that matter condensation phase transition took place [e.g., the appearance of sedentary agricultural communities], that did not constitute the transition to civilization. That represented a second transition, no longer a phase transition, but a hydrodynamic transition, a transition like the transition from laminar to turbulent flow, and for the same reason, flow convection, a

nonlinear dynamic process" (Arthur Iberall, "The Birth of Civilizations," in *The Boundaries of Civilizations in Space and Time*, eds. Matthew Melko and Leighton R. Scott [Lanham, MD: University Press of America, 1987], p. 217).

5. J. D. Becker and E. Zimmerman, "On the Dualism of Dynamics and Structure," in *The Paradigm of Self-Organization*, ed. G. J. Dalenoort (London: Gordon and Breach Science Publishers, 1989), p. 100. The authors cite a classification of self-organizing phenomena in three separate classes, according to the type (or absence) of energy flow through a system: (a) conservative (crystallization, polymerization), (b) dispersive (solitons), and (c) dissipative (chemical clocks). A good discussion of the dispersive type may be found in David Campbell, "Nonlinear Science: From Paradigms to Technicalities," in *From Cardinals to Chaos*, ed. Nacia Grant Cooper (Cambridge, UK: Cambridge University Press, 1989), p. 225. The work of Prigogine (see note 1 above) is essential to understanding the dissipative type. The mathematics of attractors and bifurcations are best explained in Ian Stewart, *Does God Play Dice? The Mathematics of Chaos* (Oxford, UK: Basil Blackwell, 1989), ch. 6.

6. The main critique of the attempt to reduce self-organization to the three types mentioned in note 5 is George Kampis, *Self-Modifying Systems in Biology and Cognitive Science* (Oxford, UK: Pergamon, 1991), ch. 5. Kampis correctly argues that even though the three orthodox types of self-organization give rise to emergent or synergistic properties, they cannot deal with *novel emergent properties*. This is particularly clear in the case of dynamical systems governed by attractors since these stable states are topological properties of phase space, and phase spaces (by definition) include *all the possible states* for a given system. It follows that (by definition) no truly novel states can be represented in phase spaces. This criticism is not fatal to those branches of nonlinear science that deal with the first three types of self-organization, since novelty and innovation are indeed *rare phenomena*. However, it does point out their limitations and calls for a new method (component systems) that can deal with novelty in terms of combinations of building blocks, and the combinatorial productivity of different blocks. This is why I use the term "nonlinear combinatorics" to designate this fourth type of self-organization. Work in this direction is also being developed at the Santa Fe Institute, as in Fontana's Turing gases or Kauffman's random grammars. See Walter Fontana, "Functional Self-Organization in Complex Systems," in Nadel and Stein, *1990 Lectures in Complex Systems*, p. 407; Stuart Kauffman, "Random Grammars," in *ibid.*, pp. 428–29.

7. See, e.g., Christopher G. Langton, "Artificial Life," in *Artificial Life*, ed. Christopher G. Langton (Redwood City, CA: Addison-Wesley, 1989). There Langton writes:

> Biology has traditionally started at the top, viewing a living organism as a complex biochemical machine, and worked *analytically* downwards from there — through organs, tissues, cells, organelles, membranes, and finally molecules — in its pursuit of the

mechanisms of life. Artificial Life starts at the bottom, viewing an organism as a large population of *simple* machines, and works upwards *synthetically* from there, constructing large aggregates of simple, rule-governed objects which interact with one another nonlinearly in the support of life-like, global dynamics. The "key" concept in Artificial Life is *emergent behavior*. Natural life emerges out of the organized interactions of a great number of nonliving molecules, with no global controller responsible for the behavior of every part.... It is this bottom-up, distributed, local determination of behavior that Artificial Life employs in its primary methodological approach to the generation of life-like behaviors. (p. 2)

8. On the new synthesis of economics and sociology (and the reasons why neoinstitutionalism is not a form of "economic imperialism"), see Viktor J. Vanberg, *Rules and Choice in Economics* (London: Routledge, 1994), ch. 1.

Viktor Vanberg and James Buchanan seem very aware of the importance of the ideas of nonequilibrium, nonlinear science for the future of both economics and sociology. These authors indeed use certain insights from Prigogine to argue for a new, nonteleological theory of markets and bureaucracies, one which assumes an *open* world of possibilities. They do not give proper emphasis, however, to the interplay between nonhuman matter-energy and human institutions, an emphasis that is necessary to truly incorporate Prigogine's insights into the study of human history. See James M. Buchanan and Viktor J. Vanberg, "The Market as a Creative Process," in *Philosophy of Economics*, ed. Daniel M. Hausman (New York: Cambridge University Press, 1994), pp. 315–28.

The concept of "transaction cost" is traced by Oliver Williamson (perhaps its best-known, although by no means only, contemporary defender) to the work of the old institutionalist school. See Oliver E. Williamson, "Transaction Cost Economics and Organization Theory," in *Organization Theory*, ed. Oliver E. Williamson (New York: Oxford University Press, 1995), pp. 207–11.

Evaluations of the research program of neoinstitutionalism from the point of view of the philosophy of science may be found in Uskali Maki, "Economics with Institutions: Agenda for Methodological Enquiry," in *Rationality, Institutions and Economic Methodology*, eds. Uskali Maki, Bo Gustafsson, and Christian Knudsen (London: Routledge, 1993), and, in the same volume, Christian Knudsen, "Modelling Rationality, Institutions and Processes."

For a more general review of method which includes also the work of "old institutionalists" (the followers of Veblen and Commons), see William Dugger, "Methodological Differences between Institutional and Neoclassical Economics," in Hausman, *Philosophy of Economics*, pp. 336–43.

9. Such a synthesis is hinted at in Robert Crosby, "Asking Better Questions," in *Cities and Regions as Nonlinear Decision Systems*, ed. Robert Crosby (Washington, DC: AAAS, 1983), pp. 9–12.

CHAPTER ONE: LAVAS AND MAGMAS

1. See Fernand Braudel, *Capitalism and Material Life, 1400–1800* (New York: Harper and Row, 1973). Braudel writes: "Geography in conjunction with the speed — or rather the slowness — of transport at the time also accounts for the very many small towns. . . . So true was it that every town welcomed movement, recreated it, scattered people and goods in order to gather new goods and new people, and so on. *It was this movement in and out of its walls that indicated the true town*" (p. 389; emphasis added).

See also Gilles Deleuze and Félix Guattari, *A Thousand Plateaus* (Minneapolis: University of Minnesota Press, 1987). Deleuze and Guattari remark: "The town is the correlate of the road. The town exists only as a function of circulation, and of circuits; it is a remarkable point [a singularity] on the circuits that create it, and that it creates. It is defined by entries and exits; something must enter it and exit from it. It imposes a frequency. It effects a polarization of matter, inert, living or human; it causes the phylum, the flow, to pass through specific places, along horizontal lines" (p. 432).

2. Indeed, the mineralizations that created our endo- and exoskeletons were bifurcations triggered by two great intensifications in the flow of energy. The first one occurred when novel forms of energy storage were "discovered" by organic evolution. New molecules called *phosphagens* allowed for the immediate provision of energy to excitable tissues (muscle and nerve), a necessary step in the development of multicellular motility. It was this flow of energy, further intensified by improvements in "phosphagen technology," that made the use of bone as a control element viable. See Ronald F. Fox, *Energy and the Evolution of Life* (New York: W.H. Freeman, 1988), pp. 94–100.

3. Richard Newbold Adams, *The Eighth Day: Social Evolution as the Self-Organization of Energy* (Austin: University of Texas Press, 1988), pp. 102–105.

4. Robert Carneiro, "Further Reflections on Resource Concentration and Its Role in the Rise of the State," in *Hunters in Transition: Mesolithic Societies of Temperate Eurasia and Their Transition to Farming*, ed. Marek Zvelebil (London: Cambridge University Press, 1986), pp. 250–51.

5. Lynn White, Jr., "The Life of the Silent Majority," in *Medieval Religion and Technology* (Berkeley: University of California Press, 1978), pp. 137–42.

6. Spiro Kostof, *The City Shaped: Urban Patterns and Meanings Through History* (London: Bulfinch, 1991), p. 30.

7. Paul M. Hohenberg and Lynn Hollen Lees, *The Making of Urban Europe, 1000–1950* (Cambridge, MA: Harvard University Press, 1985), p. 101.

8. Kostof, *The City Shaped*, pp. 46–47.

9. *Ibid.*, p. 103.

10. I use the term "distribution system" here in a very loose way to designate any institutional arrangement that affects the flow or allocation of matter-energy

resources in a given society. Karl Polanyi classified three types or modes of social integration: allocating resources via redistribution, reciprocity, and exchange. These three forms of integration are defined in terms of diagrams that plot the institutionalized pattern of flow of resources in a given society. If the pattern has a center, it is a "redistribution system"; if it is symmetrical, it is a system of reciprocity; and if it connects random points, a system of market exchange. I do not subscribe to Polanyi's theory and therefore simply adopt the idea that there are diagrams that define these flow patterns. Despite his insistence that his views are "value free" and "objective," Polanyi clearly views markets in a negative light (based on selfish gain, with an invidious element that militates against social solidarity) and views centralized regimes in a positive light. See Karl Polanyi, "Forms of Integration and Supporting Structure," in *The Livelihood of Man: Studies in Social Discontinuity*, ed. Harry W. Pearson (New York: Academic, 1972), pp. 35–61.

Braudel has severely criticized Polanyi's typology for its almost "total indifference to history" and for its "almost theological taste for definition." See Fernand Braudel, *The Wheels of Commerce* (New York: Harper and Row, 1983), pp. 225–28.

11.   Peter Sawyer, "Early Fairs and Markets in England and Scandinavia," in *The Market in History*, eds. B. L. Anderson and A. J. H. Latham (London: Croom Helm, 1986), pp. 62–64.

I must stress that I use the word "market" primarily to refer to weekly (or otherwise periodic) assemblages of people at a particular place in town. The reason for this is that, as Braudel emphasizes, it is only in these conditions that there is enough "transparency" for the participants to perceive supply and demand conditions and, hence, for prices to set themselves. The moment consumers become dispersed and linked only through chains of middlemen, this transparency is lost. Braudel still thinks that self-regulation may occur there (since prices do oscillate in unison over these larger and dispersed markets), but the exact dynamics that operate still need to be elucidated (perhaps via bottom-up simulations). See Braudel's discussion of "transparency" in *The Wheels of Commerce*, pp. 28–47; on the complexity of networks of middlemen, see *ibid.*, pp. 147–68.

Perhaps the best way to characterize the difference between "markets" as localized places in town and "markets" as dispersed sets of consumers is by going beyond the notion of "exchange" into that of "transaction" with its associated "transaction costs" (which increase as dispersion increases and information becomes harder to obtain). (See my explanation of these terms in the main text and in the following reference.)

There is another crucial difference between the two types of markets: in the case of the weekly marketplace, the entire dynamics may be disaggregated into a multiplicity of *dyadic transactions*, while a more dispersed set of consumers may give rise to more complex *network effects*. The exemplary case (in modern times) is that of the "battle" between VHS and Beta videotape formats. Although Beta was generally acknowledged to be superior on purely technical grounds, VHS won the battle due to self-reinforcing dynamics: any small advantage accumulated by one format early on in the competition was amplified by "network effects" (in this case, video-rental stores stocking more movies in VHS), leading to the entire industry becoming locked in one standard. This phenomenon (known as "path dependence") is widespread in the history of technology and has become one of the ways in which actual history is introduced in neoinstitutionalist and nonlinear economics. See, for example, Brian Arthur, "Self-Reinforcing Mechanisms in Economics," in *The Economy as an Evolving Complex System*, eds. Philip Anderson, Kenneth Arrow, and David Pines (Redwood City, CA: Addison-Wesley, 1988), pp. 10–11.

For the idea that "invisible hand" economics simply assumes that supply and demand cancel each other out (i.e., that markets clear) without ever specifying the dynamics that lead to this state, see Philip Mirowsky, *More Heat Than Light: Economics as Social Physics, Physics as Nature's Economics* (New York: Cambridge University Press, 1991), pp. 238–41. Mirowsky shows how the concept of the "invisible hand" was formalized in the nineteenth century by simply *copying the form* of equilibrium thermodynamics. (Hence, in his opinion, this branch of physics provided more heat than light.) Elsewhere he warns that recent attempts to apply Ilya Prigogine's theories to economics are making the same mistake—for example, assuming the existence of attractors without specifying just what it is that is being dissipated (i.e., only energetically dissipative or "lossy" systems have attractors). See Philip Mirowsky, "From Mandelbrot to Chaos in Economic Theory," *Southern Economic Journal* 57 (October 1990), p. 302.

12. Viktor J. Vanberg, *Rules and Choice in Economics* (London: Routledge, 1994), pp. 153–55. Karl Marx was perhaps the first to see the important connection between economic activity and social institutions (his "relations of production"). He was also the first to relate these two and the world of technology (his "means of production"). However, there are at least two things that prevent me from using Marxist concepts in this book: the labor theory of value (which Piero Schraffa has clearly shown to be a redundant part of Marxist economic theory, a kind of fifth wheel) and the built-in teleology in the traditional Marxist periodization of history as a progressive succession of modes of production (feudalism-capitalism-socialism). I believe that the elements exist today to carry out Marx's original project in a way that avoids these and other problems. The ideas expressed in this chapter are an attempt to chart this new territory, though clearly a very preliminary one.

13. *Ibid.*, pp. 127–38. Vanberg compares his own "constitutional" approach to the question of corporate actors to a prevalent sociological approach (based on the notion that having goals is what gives organizations coherence) and economic approach (based on the notion that exchanges of inducements and contributions are what gives organizations their coherence). I believe that Vanberg's solution, combining methodological individualism and ontological holism via rule-guided deci-

sion making, is the most compatible with the philosophical stance taken in this book and the most coherent means of avoiding the "functionalist" fallacy, according to which certain institutions exist because they serve the needs of an organization or of society.

A review of the history of the "goal" approach to organizations, which reveals its dependence on the "organism" metaphor, may be found in John Hassard, *Sociology and Organization Theory: Positivism, Paradigms, and Postmodernity* (New York: Cambridge University Press, 1993), chs. 1 and 2.

14. Braudel, *The Wheels of Commerce*, p. 91.

15. Brian Tierney, *The Crisis of Church and State, 1050–1300* (Toronto: University of Toronto Press, 1988), p. 7.

16. A. R. Bridbury, "Markets and Freedom in the Middle Ages," in Anderson and Latham, *The Market in History*, p. 108.

17. Hohenberg and Lees, *The Making of Urban Europe*, pp. 51–53.

18. *Ibid.*, p. 54.

19. Braudel, *Capitalism and Material Life*, pp. 394–95.

20. William H. McNeill, *The Pursuit of Power: Technology, Armed Force, and Society since A.D. 1000* (Chicago: University of Chicago Press, 1982), p. 49.

21. Braudel, *Capitalism and Material Life*, pp. 396–97.

22. White, "The Life of the Silent Majority," p. 144.

23. Howard T. Odum and Elizabeth C. Odum, *Energy Basis for Man and Nature* (New York: McGraw-Hill, 1981), p. 41.

24. Richard Hodges, *Primitive and Peasant Markets* (Oxford, UK: Basil Blackwell, 1988), p. 102. See also Deleuze and Guattari, *A Thousand Plateaus*, p. 442.

25. Hohenberg and Lees, *The Making of Urban Europe*, pp. 47–48.

26. Braudel, *Capitalism and Material Life*, p. 332.

27. William Wiseley, *A Tool of Power: The Political History of Money* (New York: John Wiley and Sons, 1977), pp. 3–4.

28. Braudel, *Capitalism and Material Life*, p. 329.

29. *Ibid.*, pp. 351 and 354–56.

30. On the role of rationality and thrift, see Braudel, *The Wheels of Commerce*, pp. 572–80.

31. Douglas C. North, *Institutions, Institutional Change, and Economic Performance* (New York: Cambridge University Press, 1995), pp. 120–31. North describes this institutional evolution as occurring along three main fronts: those that increased the mobility of capital (credit institutions); those that lowered information-acquisition costs (the printing of prices and exchange rates, the standardization of units of measure); and those that allowed the transformation of uncertainty into risk and for this risk to be spread among several agents (insurance schemes).

32. *Ibid.*, p. 127.

33. Gateway cities have played a key role in history since ancient times and

coexisted with both "primitive" and state societies, supplying their elites with luxury items. See Hodges, *Primitive and Peasant Markets*, pp. 42–51.

34. Hohenberg and Lees, *The Making of Urban Europe*, p. 240.

35. *Ibid.*, p. 64.

36. Fernand Braudel, *The Perspective of the World* (New York: Harper and Row, 1986), pp. 27–31.

37. Hohenberg and Lees, *The Making of Urban Europe*, p. 165. The terminology "core," "semiperiphery," and "periphery" are typically associated with Immanuel Wallerstein's highly regarded theory of world-economies. These are the largest units of analysis in economic history, large areas of economic coherence involving trans-national trade networks and hence encompassing territories larger than countries (although not necessarily of planetary proportions, like today's world-economy). Although I acknowledge the importance of Wallerstein's work as a contribution to the empirical study of emergent structures at this scale, the teleology involved in his theory of stages (even though it is an improvement over the linear sequence feudalism-capitalism-socialism) and his intensified methodological holism (now tak-ing as its point of departure for a top-down analysis a much larger entity than a single society) prevent me from using his theories in this book. See, for example, his stance on "stages" and the need to begin one's study with the largest "totali-ties," in "The Rise and Future Demise of the World Capitalist System: Concepts for Comparative Analysis," in Immanuel Wallerstein, *The Capitalist World-Economy* (New York: Cambridge University Press, 1993).

Fortunately, Wallerstein's approach is not the only one available. Braudel has developed an alternative theory of world-economies that is (at least potentially) very valuable for a more bottom-up approach. As I said before, the idea here is to com-bine methodological individualism and ontological holism. That is, to start at the bottom with individual decision makers and transactors and then derive subsequent entities on larger scales (institutional organizations, cities, states, world-economies) one layer at a time. Hence, this approach shares Wallerstein's ontological holism (i.e., the idea that these larger entities have an autonomous existence in reality) but not his top-down methodology. Braudel's approach seems an intermediate one. His main source of disagreement with Wallerstein is over the temporal and spatial limits of world-economies. While for Wallerstein only Europe gave rise to this phe-nomenon (other areas of the world, such as China or Islam, created world empires instead), for Braudel these areas had world-economies as real and powerful as those of Europe, although with some major differences, such as the absence of antimar-kets and the presence of a semiperiphery: "From earliest times, the core or 'heart' of Europe was surrounded by a nearby semi-periphery and by an outer periphery. And the semi-periphery, a pericardium so to speak enclosing the heart and forcing it to beat faster — northern Italy around Venice in the fourteenth and fifteenth centuries, the Netherlands around Antwerp — was probably the essential feature of the structure of Europe. There does not seem to have been a semi-periphery around Peking or Delhi, Isfahan, Istanbul or Moscow" (Braudel, *The Perspective of the World*, p. 56).

Besides disagreeing on the spatial distribution of world-economies in history, these authors also differ in the temporal limits of these large-scale entities. For Wallerstein the European world-economy begins in the sixteenth century, with the failure of the Hapsburg Empire to create a world empire. This is, of course, necessary for his argument, since he needs to convince us that there has been only one world-economy, and that that world-economy may be identified with "capitalism." However, Braudel disagrees:

> I am therefore inclined to see the European world-economy as having taken shape very early on; I do not share with Immanuel Wallerstein's fascination with the sixteenth century.... For Wallerstein, the European world-economy was the matrix of capitalism. I do not dispute this point since to say central zone [i.e., what I refer to here as "core of the Network system"] or capitalism is to talk about the same reality. By the same token, however, to argue that the world-economy built in the sixteenth century on its European site was not the first to occupy this small but extraordinary continent, amounts to saying that capitalism did not wait until the sixteenth century to make its first appearance. I am therefore in agreement with the Marx who wrote (although he later went back on this) that European capitalism — indeed he even says capitalist *production* — began in thirteenth-century Italy. This debate is anything but academic. (*ibid.*, p. 57)

Clearly, Braudel himself does not completely reject Marxist approaches to this question. (See his discussion and criticism of Wallerstein's concept, in *ibid.*, pp. 51–57.) I feel more inclined to start the analysis of world-economies from scratch and bottom-up, using nonlinear models to explain their temporal coherence (e.g., economic waves of different durations) and urban dynamics (e.g., the analysis of the Network system of Hohenberg and Lees) to account for its spatial coherence. This seems to me the only way to get rid of teleological (or stagelike) accounts of history still very evident in Marxist terms such as "late capitalism." I am aware, however, that such a sketchy account as I have been able to give here will hardly seem convincing to anyone who already operates within the Wallersteinian paradigm. A serious account of this and other related issues will have to wait for another occasion.

38. Hohenberg and Lees, *The Making of Urban Europe*, p. 281.

39. *Ibid.*, p. 282.

40. Descriptions and criticisms of Christaller's theory may be found in *ibid.*, pp. 49–50, and Hodges, *Primitive and Peasant Markets*, pp. 16–34.

41. Dimitrios Dendrinos, *Urban Evolution* (Oxford, UK: Oxford University Press, 1985), pp. 31 and 45–46; Peter M. Allen, "Self-Organization in the Urban System,"

NOTES

in *Self-Organization and Dissipative Structures: Applications in the Physical and Social Sciences*, eds. William C. Schieve and Peter M. Allen (Austin: University of Texas, 1982), pp. 135–36; Peter M. Allen, "Self-Organization and Evolution in Urban Systems," in *Cities and Regions as Nonlinear Decision Systems*, ed. Robert Crosby (Washington, DC: AAAS, 1983), pp. 39–45.

    42. Herbert Simon, *The Sciences of the Artificial* (Cambridge, MA: MIT Press, 1994), pp. 32–36.

    43. Richard Day, "Adaptive Economics," in Crosby, *Cities and Regions as Nonlinear Decision Systems*, pp. 103–39; Richard Day, "The General Theory of Disequilibrium Economics and of Economic Evolution," in *Economic Evolution and Structural Adjustment*, eds. D. Batten, J. Casti, and B. Johansson (Berlin: Springer Verlag, 1987), pp. 46–61; Siro Lombardini, "Rationality in Disequilibrium," in *Nonlinear and Multisectoral Macrodynamics*, ed. Kumaraswamy Velupillai (New York: New York University Press, 1990), pp. 207–22.

For historical evidence that "skills" and not some general "rationality" are what drove decision making in the early centuries of the modern European economy, and that these skills needed to be learned via a system of apprenticeship (sending sons to trading posts), see Braudel, *The Wheels of Commerce*, pp. 405–408.

    44. This is a well-known result in nonlinear economics at least since the work of Richard Goodwin in the 1940s and 1950s. See, for example, Richard M. Goodwin, "On Growth and Form in an Economy," in *Essays in Nonlinear Economic Dynamics* (Frankfurt: Verlag Peter Lang, 1989), p. 24. See also remarks on self-regulation and nonoptimization, in Simon, *The Sciences of the Artificial*, p. 43.

    45. Braudel, *The Wheels of Commerce*, pp. 227–28, and *The Perspective of the World*, pp. 71–87.

    46. That data from several economic indicators (GNP, unemployment rate, aggregate prices, interest rates), beginning in the early nineteenth century, display an unequivocal periodic motion of approximately fifty years' duration (called Kondratieff cycles) is well known at least since the work of Joseph Schumpeter. Several possible mechanisms to explain this cyclical behavior have been offered since then, but none has gained complete acceptance (most of the models are top-down). A bottom-up MIT model endogenously generates this periodic oscillation, with the behavior emerging spontaneously from the interaction of different segments of the population of organizations, as well as nonlinearities (such as delays). See Jay W. Forrester, "Innovation and Economic Change," in *Long Waves in the World Economy*, ed. Christopher Freeman (Boston: Butterworth, 1983), p. 128. (This volume also offers a survey of the different theories of the long wave.) On the MIT model and the constructive role that delays may play, see J. D. Sterman, "Nonlinear Dynamics in the World Economy: The Economic Long Wave," in *Structure, Coherence and Chaos in Dynamical Systems*, eds. Peter L. Christiansen and R. D. Parmentier (Manchester, UK: Manchester University Press, 1989).

47. Thomas F. Glick, "Science, Technology and the Urban Environment: The Great Stink of 1858," in *Historical Ecology*, ed. Lester J. Bilsky (New York: Kennikat, 1980), p. 128. A more general theory of bureaucracies in modern democratic societies which shows the processes through which their efficiency is constantly compromised by questions of power and power struggles may be found in Terry M. Moe, "The Politics of Structural Choice: Toward a Theory of Public Bureaucracy," in *Organization Theory*, ed. Oliver E. Williamson (New York: Oxford University Press, 1995), pp. 116–49.

48. There are several exceptions to this "rule" (individual decision making affects only one level of scale). One of them involves special situations where the level immediately higher (the level of institutions) is near a bifurcation point in *its own dynamics*. Here the decisions and actions of individuals may be amplified and have effects beyond their scale. I have often quoted Prigogine and Isabelle Stengers on this point:

> From the physicist's point of view this involves a distinction between states of the system in which all individual initiative is doomed to insignificance on one hand, and on the other, bifurcation regions in which an individual, an idea, or a new behavior can upset the global state. Even in those regions, amplification obviously does not occur with just any individual, idea, or behavior, but only with those that are "dangerous" — that is, those that can exploit to their advantage the nonlinear relations guaranteeing the stability of the preceding regime. Thus we are led to conclude that the same nonlinearities may produce an order out of the chaos of elementary processes and still, under different circumstances, be responsible for the destruction of this same order, eventually producing a new coherence beyond another bifurcation. (Ilya Prigogine and Isabelle Stengers, *Order Out of Chaos: Man's New Dialogue with Nature* [New York: Bantam, 1984], p. 190)

49. Braudel, *The Wheels of Commerce*, p. 315.

50. Jane Jacobs, *Cities and the Wealth of Nations* (New York: Random House, 1984), p. 40.

51. *Ibid.*, p. 50.

52. *Ibid.*, p. 144.

53. Norman H. Packard, "Dynamics of Development: A Simple Model for Dynamics Away from Attractors," in Anderson et al., *The Economy as an Evolving Complex System*, p. 175. In the same volume, other properties of economic meshworks are explored in Stuart A. Kauffman, "The Evolution of Economic Webs," and John H. Holland, "The Global Economy as an Adaptive Process." Besides their insights on meshwork dynamics, these essays illustrate three different approaches to "nonlinear combinatorics," that is, dynamics *away from global attractors*.

54. Jacobs, *Cities and the Wealth of Nations*, p. 43.

55. Braudel, *The Wheels of Commerce*, p. 379.

56. Braudel, *The Perspective of the World*, p. 630; and John Kenneth Galbraith, *The New Industrial State* (Boston: Houghton Mifflin, 1978), p. xvii.

57. Braudel, *The Perspective of the World*, p. 631.

58. The type of competition in which oligopolies engage is of the type studied by game theory. Here every actor (large firm) must take into account the potential reactions of other actors to each one of its moves, as it plans new strategies for future action. (For example, a large firm cannot unilaterally lower prices without fear of triggering a price war.) In a real market, however, there are so many actors that no one can plan future courses of action which take into account every potential rival. See John R. Munkirs and James I. Sturgeon, "Oligopolistic Cooperation: Conceptual and Empirical Evidence of Market Structure Evolution," in *The Economy as a System of Power*, eds. Marc R. Tool and Warren J. Samuels (New Brunswick, NJ: Transaction, 1989), p. 338.

But beyond this difference, the main feature distinguishing self-regulating markets from oligopolistic competition is that the actors involved in the former are *price takers*, that is, they have no control whatsoever over price determination, which is basically automatic. The latter, on the other hand, are *price makers*, since they establish their own prices by some heuristic procedure, such as adding a markup to the costs of production. Orthodox economists, accepting that oligopolies set their own prices, attempt to rescue their theory by asserting that the price big corporations arrive at is the one that maximizes their profits, and since the optimality of this price is objectively determined by outside forces, in a sense, this price is still setting itself. For a nonorthodox reply and for the history of this controversy see, for example, Dennis C. Mueller, "The Corporation and the Economist," in *Philosophy of Economics*, ed. Daniel M. Hausman (New York: Cambridge University Press, 1994), pp. 293–98.

Orthodox economists have found an equivalent to the "invisible hand" in oligopolistic competition: Nash equilibriums, defined as a set of strategies, one for each player, such that no player can improve his expected utility by unilaterally changing his strategy. However, as Mario Henrique Simonsen shows, this ideal outcome may be, for a variety of reasons, impossible to achieve. (Prudence, for example, on the part of one of the competing oligopolistic firms, may get in the way. To achieve a Nash state all firms must gamble on the assumption that the rest are shooting for a Nash strategy.) Only the "visible hand" of government intervention (in the form of old-fashioned Keynesian management of aggregate demand) can solve this. See Mario Henrique Simonsen, "Rational Expectations, Game Theory and Inflationary Inertia," in Anderson et al., *The Economy as an Evolving Complex System*, pp. 205–208.

Market power, in its different manifestations, seems also the weak point of the neoinstitutionalist economics on which I have relied up to this point. On the other hand, "old" institutional economists (e.g., the modern-day followers of Veblen) have never lost sight of power. As they point out, there are several ways in which oligopolistic competition may be turned into cooperation, which in this case is not something to be valued positively since it is equivalent to a monopolistic state of affairs. (If Ford, GM, and Chrysler "cooperate" in setting prices, for example, they become one big monopoly.) One way in which this transformation may occur is if the board of directors of each firm includes members of the same banking or insurance companies. This is the phenomenon known as "interlocking directorates" for which much indirect evidence exists. (Such a structure would be, in effect, a meshwork of hierarchies, in my terminology.) See John Murkirs, "Centralized Private Sector Planning: An Institutionalist Perspective on the Contemporary U.S. Economy," in Tool and Samuels, *The Economy as a System of Power*, pp. 285–96.

Despite the absolute necessity of distinguishing between the market-theoretic and the game-theoretic meanings of the word "competition," the distinction in practice cannot be applied in this strongly dichotomized way, since it is clear that some small firms grow into large ones, that even oligopolies still deal with markets for some of their inputs, and so on. Hence the need to stress the idea of "complex mixtures," a dynamics that may not be analytically tractable and hence may need bottom-up simulations to be studied.

59. Braudel, *The Perspective of the World*, pp. 103–104 and 124–28.

60. Braudel writes, "I am tempted to agree with Deleuze and Guattari that 'after a fashion, capitalism has been a spectre haunting every form of society' — capitalism, that is, as I have defined it [i.e., as antimarkets]" (*ibid.*, p. 581).

61. *Ibid.*, p. 559.

62. On the analysis of non-European world-economies, see Braudel, *The Perspective of the World*, pp. 523–29. See also note 37 above.

63. Ad hoc redefinition of terms is one of the strategies that may immunize a theory against falsification, according to Popperian philosophy of science. However, one does not have to be a strict Popperian (i.e., to see falsifiability as *the* landmark of scientific knowledge) to realize the dangers involved in ad hoc redefinitions. On the virtues and limitations of Popper's and Lakatos's approaches when applied to economics, see, for example, Mark Blaug, "Why I Am Not a Constructivist: Confessions of an Unrepentant Popperian," in *New Directions in Economic Methodology*, ed. Roger E. Backhouse (London: Routledge, 1994), pp. 109–15.

Braudel himself prefers to keep the word "capitalism" and change its meaning (so that it refers exclusively to non-market-competition, i.e., big business). However, such an entrenched meaning cannot be changed so easily. This is why I prefer to use a different term altogether, and one which bears its intended meaning on its sleeve. A term like "antimarket" is precisely what is needed here to wrest the notion of "market" both from the right (invisible handers) and the left (commodifiers). This, it seems to me, is a crucial move, otherwise we will be confined, when thinking about possible routes for social development, between two choices that are equally hierarchical: capitalism and socialism. On the history of the word "capitalism," see Braudel, *The Wheels of Commerce*, pp. 232–38.

64. Braudel, *The Wheels of Commerce*, p. 419.

65. *Ibid.*, p. 405.

66. *Ibid.*, pp. 97–100 and 390–95. Bills of exchange and other forms of primitive paper money, such as banknotes, arose more or less spontaneously out of the daily activities of big merchants, and whenever metallic money was not plentiful enough to catalyze trade. Similarly, banks and stock exchanges emerged first as informal practices, becoming institutions as the rules that governed them hardened into formal procedures. Only later on did these institutional practices became "mineralized," as banks and exchanges acquired their own permanent buildings. For example, stocks on government loans circulated through the top of commercial hierarchies (i.e., big fairs) as early as the fourteenth century. Early stock exchanges were like the upper echelons of fairs, only operating permanently, originally simply as daily meetings of wealthy merchants and brokers at a given spot in many medieval cities. By the time special buildings were built to house these meetings, they had already developed formal rules for conducting their transactions. Thus, while the exchange at Antwerp was in existence by 1460, its mineralization did not occur until 1518. And a similar point can be made about banks, which emerged as dispersed practices, whether of money lenders or the services that merchant companies performed for one another, later evolving into separate institutions in Florence around the fourteenth century. A banking system, however, would take longer to develop and cannot be said to have been in place until the eighteenth century, centered in Amsterdam, the core of the Network system at the time.

On the banking system's difficulty in establishing itself, and on the contingent history of banks and banking, which does not reflect any underlying rationality, see John Kenneth Galbraith, *Money: Whence It Came, Where It Went* (Boston: Houghton Mifflin, 1975), chs. 3–8.

67. Anne Querrien, "The Metropolis and the Capital," in *Zone 1/2: The Contemporary City* (New York: Zone, 1986), p. 219.

68. Hohenberg and Lees, *The Making of Urban Europe*, p. 70.

69. Paul Kennedy, *The Rise and Fall of the Great Powers* (New York: Random House, 1987), pp. 22–23.

70. Braudel, *Capitalism and Material Life*, p. 386.

71. Kennedy, *The Rise and Fall of the Great Powers*, pp. 11–12.

72. McNeill, *The Pursuit of Power*, p. 44.

73. *Ibid.*, p. 45. On China's missed opportunity to "discover" Europe, see also Braudel, *The Wheels of Commerce*, p. 581, and Kennedy, *The Rise and Fall of the Great Powers*, p. 7.

74. Braudel, *The Perspective of the World*, p. 32.

75. Alfred W. Crosby, *Ecological Imperialism: The Biological Expansion of Europe, 900–1900* (New York: Cambridge University Press, 1989), p. 107.

76. *Ibid.*, pp. 113–14.

77. Hohenberg and Lees, *The Making of Urban Europe*, p. 161.

78. City hierarchies interacted in ways that promoted autocatalytic dynamics. Urban antimarkets, for example, provided credit to finance the wars that defeated every effort to make Europe into a homogeneous hierarchy (for example, Amsterdam's financiers supplied the funds that London needed to defeat Napoleon and thus keep the continent a heterogeneous meshwork). On the financial aspects of war, and the differences between France and England in this respect, see Kennedy, *The Rise and Fall of the Great Powers*, pp. 80–85.

79. Malakondavya Challa and Richard L. Pfeffer, "Formation of Atlantic Hurricanes from Cloud Clusters and Depressions," in *Journal of Atmospheric Sciences* (April 1, 1990), p. 909.

80. Harvey Blatt, Gerard Middleton, and Raymond Murray, *Origin of Sedimentary Rocks* (New York: Prentice-Hall, 1972), p. 102.

81. *Ibid.*, p. 353.

82. Deleuze and Guattari, *A Thousand Plateaus*, p. 41. Deleuze and Guattari call these two operations "content" and "expression" and warn us against confusing them with the old philosophical distinction between substance and form. Content and expression each involves substance and form: sedimentation is not just about accumulating pebbles (substance) but also about sorting them into uniform layers (form); while consolidation not only effects new architectonic couplings between pebbles (form) but also yields a new entity, a sedimentary rock (substance). It is this form of the diagram (one operation involving substances and forms, another operation involving forms and substances) that is the most abstract and, hence, the most useful. The particular instantiation that I will be using in this book (sorting + consolidation) may be seen as a particular form of this more general diagram.

Actually, here Deleuze and Guattari incorrectly characterize the two articulations involved in rock production as "sedimentation-folding." The correct sequence is "sedimentation-cementation." Then, *on a different spatial scale*, "cyclic sedimentary rock accumulation-folding into mountain." In other words, they collapse two different double-articulations (one utilizing as its starting point the products of the previous one) into one. I believe this correction does not affect their underlying argument and indeed strengthens it (since it shows that the same process may occur on two different scales).

83. Niles Eldridge, *Macroevolutionary Dynamics: Species, Niches, and Adaptive Peaks* (New York: McGraw-Hill, 1989), p. 127.

84. Marvin Harris, *Cannibals and Kings* (New York: Vintage, 1991), p. 104.

85. S. N. Eisenstadt, "Continuities and Changes in Systems of Stratification," in

Stability and Social Change, eds. Bernard Barber and Alex Inkeles (Boston: Little, Brown, 1971), p. 65.

86. *Ibid.*, pp. 66–71.

87. Humberto R. Maturana and Francisco J. Varela, *The Tree of Knowledge: The Biological Roots of Human Understanding* (Boston: Shambhala, 1992), pp. 47 and 115. Other researchers have discovered that as the loop adds new nodes it may reach a critical threshold of complexity and undergo a bifurcation, a transition to a new state where complexification accelerates. (What I referred to above as "industrial takeoff.") Since the states to which a phase transition leads are in no way "directed" or "progressive," changing and developing by crossing bifurcations are other ways of growing by drift.

88. Prigogine and Stengers, *Order Out of Chaos*, p. 147.

89. Francisco J. Varela, "Two Principles of Self-Organization," in *Self-Organization and Management of Social Systems*, eds. H. Ulrich and G. J. B. Probst (Berlin: Springer Verlag, 1984), p. 27.

90. Deleuze and Guattari, *A Thousand Plateaus*, p. 329.

91. Michael Bisacre, *Encyclopedia of the Earth's Resources* (New York: Exeter, 1984), p. 79.

92. Deleuze and Guattari, *A Thousand Plateaus*, p. 328. The authors constantly refer to catalysis in their theories of meshworklike structures (rhizomes, smooth spaces, etc.). They tend to view catalysis in terms of one specific (albeit very important) type of catalyst: the allosteric enzymes discovered by Jaques Monod, which are like programmable catalysts, with two heads. "What holds heterogeneities together without their ceasing to be heterogeneous . . . are intercalary oscillations, synthesizers with at least two heads" (p. 329).

What is needed here is to make the notion of a "catalyst" more abstract so that the specific functions of a chemical catalyst (to perform acts of recognition via a lock-and-key mechanism, to accelerate or decelerate chemical reactions) are not what matters, but the more general notion of aiding growth "from within" or "from in between." One step in this direction has been taken by Arthur Iberall, whom I mentioned in the introduction as a pioneer in "nonlinear history," by defining catalytic activity as the ability to force a dynamical system from one attractor to another. In the case of a chemical catalyst the dynamical system would be the target molecule (the one to be catalyzed) and the two stable states would be its "unreactive" and "reactive" states, so that by switching molecules from one state to another the catalyst accelerates the reaction. See Arthur Iberall and Harry Soodak, "A Physics for Complex Systems," in *Self-Organizing Systems: The Emergence of Order*, ed. Eugene Yates (New York: Plenum, 1987), p. 509.

Elsewhere, Iberall notes that, in this sense, nucleation events and dislocations may be considered to involve "acts of catalysis." *Nucleation* refers to the process through which the structures that appear after a phase transition (crystals just after the bifurcation to the solid state, for example) consolidate and grow, as opposed to reverting back to the previous state (by crossing the bifurcation in the opposite direction). Typically, something has to catalyze the growth of structure to a critical mass (nucleation threshold), after which growth may proceed more or less spontaneously. This "something" may be anything from a dust particle to a defect in the container in which the crystallization is happening. If one carefully removes all particles and defects, one can indeed cool down a liquid past the bifurcation point without crystallization taking place. (Eventually, as we cool down further, even a microscopic thermal fluctuation can act as catalyst and trigger the nucleation.) *Dislocations*, on the other hand, are line defects within the body of the growing crystals which help them grow by storing mechanical energy in their misaligned (hence nonequilibrium) composing atoms. This stored energy allows them to promote crystal growth by lowering nucleation thresholds. Thus, in this abstract sense of "catalysis," the intercalary events involved in the creation of igneous rocks are of the meshwork-generating type. On this see Arthur Iberall, *Toward a General Science of Viable Systems* (New York: McGraw-Hill, 1972), p. 208.

But we can go further. Defined this way, "catalysis" becomes a true abstract operation: anything that switches a dynamical system (an interacting population of molecules, ants, humans, or institutions) from one stable state to another is *literally* a catalyst in this sense. Hence, we may use this definition not only to move down from chemistry (the field of the literal application of the term) to physics, without metaphor, but also up, to biology, sociology, and linguistics. In this book I will use the term to refer to this abstract operator capable of constraining matter-energy flows of different kinds, by switching them from attractor to attractor. Cities and institutions, for example, would be instantiations of this operator to the extent that they arise form matter-energy flows and decision-making processes but then react back on these flows and processes to constrain them in a variety of ways (stimulating them or inhibiting them). On the other hand, as Iberall himself notes, catalytic constraints may combine with one another and form languagelike systems. Another physicist, Howard Pattee, has further elaborated the notion of enzymes (organic catalysts) as syntactical constraints, operating on a semantic world defined by its stable states. This will be important in Chapter Three, where I will discuss a recent mathematical theory of language (by Zellig Harris) based precisely on the notion of combinatorial constraint (which replaces that of "grammatical rule"). On biological catalysts as syntactic constraints, see Howard Pattee, "Instabilities and Information in Biological Self-Organization," in Yates, *Self-Organizing Systems*, p. 334.

93. Gregoire Nicolis and Ilya Prigogine, *Exploring Complexity* (New York: W.H. Freeman, 1989), p. 29.

94. Deleuze and Guattari, *A Thousand Plateaus*, p. 335.

95. See, for example, Russell D. Vetter, "Symbiosis and the Evolution of Novel Trophic Strategies," in *Symbiosis as a Source of Evolutionary Innovation*, eds. Lynn

Margulis and Rene Fester (Cambridge, MA: MIT Press, 1991), pp. 219–40, and Peter W. Price, "The Web of Life: Development over 3.8 Billion Years of Trophic Relations," in *ibid.*, pp. 262–70.

96. In the opinion of the ecologist Stuart Pimm, interviewed in Roger Lewin, *Complexity: Life at the Edge of Chaos* (New York: Macmillan, 1992), p. 126.

97. Simon, *The Sciences of the Artificial*, p. 41.

98. North, *Institutions, Institutional Change, and Economic Performance*, p. 108. In real markets beyond a certain level of scale and complexity, primitive money and informal constraints are not enough to articulate heterogeneous demands. Monetary systems (with a strong hierarchical structure) as well as formal constraints are needed to keep transaction costs down. According to North, formal rules often form hierarchies, too, with the rules at the top of the pyramid changing very slowly and those at the bottom changing more swiftly: "Formal rules include political (and judicial) rules, economic rules, and contracts. The hierarchy of such rules, from constitutions, to statute and common laws, to specific bylaws, and finally to individual contracts, defines constraints, from general rules to particular specifications. And typically constitutions are designed to be more costly to alter than statute laws, just as statute law is more costly to alter than individual contracts" (p. 47).

99. Simon, *The Sciences of the Artificial*, p. 38.

100. As Deleuze and Guattari write:

> Stating the distinction in its more general way, we could say that it is between stratified systems or systems of stratification on the one hand, and consistent, self-consistent aggregates on the other.... There is a coded system of stratification whenever, horizontally, there are *linear causalities* between elements; and, vertically, hierarchies of order between groupings; and, holding it all together in depth, a succession of framing forms, each of which informs a substance and in turn serves as a substance for another form [e.g., the succession pebbles-sedimentary rocks-folded mountains above].... On the other hand, we may speak of aggregates of consistency when instead of a regulated succession of forms-substances we are presented with consolidations of very heterogeneous elements, orders that have been short-circuited or even *reverse causalities*, and captures between materials and forces of a different nature. (Deleuze and Guattari, *A Thousand Plateaus*, p. 335; emphasis added)

I take it that here the expression "reverse causalities" refers to circular causality or feedback mechanisms.

101. Magoroh Maruyana, "Symbiotization of Cultural Heterogeneity: Scientific, Epistemological and Aesthetic Bases," in *Cultures of the Future*, eds. Magoroh Maruyana and Arthur M. Hrakins (The Hague: Mouton, 1978), pp. 457–58; and Magoroh Maruyana, "Four Different Causal Metatypes in Biological and Social Sciences," in Schieve and Allen, *Self-Organization and Dissipative Structures*, p. 355.

102. An example of the use of the term "anti-Gaian" to refer to positive feedback is Penelope J. Boston and Starley L. Thompson, "Terrestrial Microbial and Vegetation Control of Planetary Environments," in *Scientists on Gaia*, eds. Stephen H. Schneider and Penelope J. Boston (Cambridge, MA: MIT Press, 1993), p. 99. A criticism of the use of "Gaian" terms as a mere relabeling of positive and negative feedback may be found in James W. Kirchner, "The Gaia Hypotheses: Are They Testable? Are They Useful?" in *ibid.*, p. 38.

103. Maruyama, "Symbiotization of Cultural Heterogeneity," pp. 459–60.

104. *Ibid.*, p. 470.

105. John D. Steinbruner, *The Cybernetic Theory of Decision* (Princeton, NJ: Princeton University Press, 1974), pp. 47–55. This book is about the role that negative feedback may play in institutions as a kind of homeostatic mechanism. In economics, negative feedback appears mostly in the form of "diminishing returns."

106. Michael J. Radzicki, "Institutional Dynamics, Deterministic Chaos and Self-Organizing Systems," *Journal of Economic Issues* 24 (March 1990). The author proposes a model of institutional dynamics as "a mathematical pattern of positive and negative feedback loops, containing accumulations or numerical integrations that are joined together by nonlinear couplings" (p. 59).

107. George Kampis, *Self-Modifying Systems in Biology and Cognitive Science: A New Framework for Dynamics, Information and Complexity* (Oxford, UK: Pergamon, 1991), p. 235. Kampis writes: "The notion of immensity translates as irreducible variety of the component-types.... This kind of immensity is an immediately complexity-related property, for it is about variety and heterogeneity, and not simply as numerousness."

108. Josef W. Konvitz, *Cities and the Sea: Port City Planning in Early Modern Europe* (Baltimore, MD: Johns Hopkins University Press, 1978), p. 73.

109. Hohenberg and Lees, *The Making of Urban Europe*, p. 185.

110. This idea, that there may be fortuitous accumulations of complexity but not a general drive toward complexification has been defended most eloquently by Stephen Jay Gould. See, for example, "Tires to Sandals," in Stephen Jay Gould, *Eight Little Piggies* (New York: W.W. Norton, 1994), pp. 318–24. See also the opinions expressed by Gould and Richard Dawkins, quoted in Lewin, *Complexity: Life at the Edge of Chaos*, pp. 145–46.

As far as the evolution of technology is concerned, for the idea that technological development does not follow a single line, that many possible lines are left undeveloped, and that there have always been different alternatives (some more oppressive and controlling than others), even for mass production, see Seymour Melman, "The Impact of Economics on Technology," in Tool and Samuels, *The Economy as a System of Power*, pp. 49–61.

This is a theme related to Brian Arthur's theory regarding network externalities due to positive feedback (the phenomenon is referred to as "path dependence").

One of the possibilities is always that a particular technology will become locked in the inferior standard. A similar point may apply to institutional evolution. See North, *Institutions, Institutional Change, and Economic Performance*, ch. 11.

111. Hohenberg and Lees, *The Making of Urban Europe*, p. 185.

112. *Ibid.*, p. 197.

113. Ian G. Simmons, *Changing the Face of the Earth: Culture, Environment, History* (Oxford, UK: Basil Blackwell, 1989), p. 216.

114. Hohenberg and Lees, *The Making of Urban Europe*, p. 243.

115. Simmons, *Changing the Face of the Earth*, p. 201.

116. Richard Newbold Adams, "The Emergence of Hierarchical Social Structure: The Case of Late Victorian England," in Schieve and Allen, *Self-Organization and Dissipative Structures*, p. 124.

117. Adams, *The Eighth Day*, p. 133.

118. George F. Ray, "Innovation and Long Term Growth," in Freeman, *Long Waves in the World Economy*, p. 184.

119. Braudel, *The Perspective of the World*, p. 548.

120. *Ibid.*, pp. 548–49.

121. *Ibid.*, pp. 552–53.

122. *Ibid.*, p. 560.

123. Lynn White, Jr., "Pumps and Pendula," in *Medieval Religion and Technology*, p. 130.

124. Eugene S. Ferguson, *Engineering and the Mind's Eye* (Cambridge, MA: MIT Press, 1993), pp. 58–59. On the role of information and skills in the Industrial Revolution, see Ian Inkster, *Science and Technology in History* (New Brunswick, NJ: Rutgers University Press, 1991), ch. 3.

125. Braudel, *The Perspective of the World*, pp. 277 and 294–95.

126. *Ibid.*, p. 385.

127. *Ibid.*, p. 588.

128. Carl W. Condit, "Buildings and Construction," in *Technology in Western Civilization*, 2 vols., eds. Melvin Kranzberg and Carrol W. Pursell (New York: Oxford University Press, 1967), vol. 1, pp. 374–75.

129. Hohenberg and Lees, *The Making of Urban Europe*, pp. 241–42. There, the authors write:

> Rail junctions such as Crewe and Vierzon joined river and canal ports and towns at the mouth of valleys as commercially strategic places.... The Network System in the nineteenth and twentieth centuries broke free of the constraints heretofore imposed on it by ports and strategic crossroads. Although many traditional nodes and gateways continued to flourish, the pull of territorial capitals on trade, finance and enterprise could grow unchecked. With their concentration of power and wealth, these cities commanded the design of the rail networks and later of the motorways, and so secured

the links on which future nodality depended. Where once the trade routes and water-ways had determined urban location and roles in the urban network, rail transporta-tion now accommodated the expansion needs of the great cities for both local traffic and distant connections.

130. Eugene S. Ferguson, "Steam Transportation," in Kranzberg and Pursell, *Technology in Western Civilization*, vol. 1, pp. 296–97.

131. Jacobs, *Cities and the Wealth of Nations*, p. 145. See also Braudel, *The Perspective of the World*, pp. 409–10 and 426, on the role of maritime gateways in eighteenth-century American colonies.

132. Roger Burlingame, "Locomotives, Railways, and Steamships," in Kranzberg and Pursell, *Technology in Western Civilization*, vol. 1, p. 429.

133. Charles F. O'Connell, Jr., "The Corps of Engineers and the Rise of Modern Management, 1827–1856," in *Military Enterprise: Perspectives on the American Expe-rience*, ed. Merrit Roe Smith (Cambridge, MA: MIT Press, 1987), pp. 88–89.

134. Robert C. Davis, *Shipbuilders of the Venetian Arsenal: Workers and Workplace in the Preindustrial City* (Baltimore, MD: Johns Hopkins University Press, 1991), p. 44.

135. Harry Braverman, *Labor and Monopoly Capital* (New York: Monthly Review Press, 1974), p. 89.

136. Merrit Roe Smith, "Army Ordnance and the 'American System of Manufac-turing,' 1815–1861," in Smith, *Military Enterprise*, p. 79. The classical work in this area is David A. Hounshell, *From the American System to Mass Production, 1800–1932* (Baltimore, MD: Johns Hopkins University Press, 1984), ch. 1. See also note 175 below, on the history of automation, and the discussion of this and other interactions between military and economic institutions in Manuel De Landa, *War in the Age of Intelligent Machines* (New York: Zone, 1992), ch. 1.

Recently, the purely military origin of the American system has been challenged in Donald R. Hoke, *Ingenious Yankees: The Rise of the American System of Manufac-turers in the Private Sector* (New York: Columbia University Press, 1990). However, it seems to me that Hoke's criticisms fall short. He acknowledges that the basic idea behind the system (a standard model to be copied exactly) was born in French eighteenth-century arsenals and adopted later in the U.S. through imitation—for example, by the early wooden-clock manufacturers operating on the "putting out" system (which antedated concentrated factory production). His other examples are all big business and so are not really counterexamples but simply examples of convergence toward disciplinary methods by large hierarchical organizations. To pretend that those large organizations were being driven by "Yankee ingenuity" is naive, given the hierarchical nature of those institutions.

137. Braudel, *The Wheels of Commerce*, pp. 322–25.

138. Some recent nonlinear models of economic evolution stress the interaction between two different processes, innovation and routinization—that is, between the

spontaneous proliferation of flexible skills and procedures and their gradual conversion into rigid, uniform routines. According to these models, the process of innovation pushes economic evolution far from equilibrium, toward the multiple forms of stability that characterize self-organization, while the process of routinization brings the economy back to equilibrium. See, for example, D. Batten, J. Casti, and B. Johanson, "Economic Dynamics, Evolution and Structural Adjustment," in *Economic Evolution and Structural Adjustment*, eds. D. Batten, J. Casti, and B. Johanson (Berlin: Springer Verlag, 1987), pp. 19–20.

139. Hohenberg and Lees, *The Making of Urban Europe*, p. 203.

140. *Ibid.*, p. 202. See also A. E. Anderson, "Creativity and Economic Dynamics Modelling," in Batten et al., *Economic Evolution and Structural Adjustment*, pp. 27–44. Anderson mentions several well-studied cases of "creative explosions" in urban centers due to economies of agglomeration (deep knowledge in a number of fields and intensive local interaction). In addition to this, he mentions the need for a sponsoring institution and a perceived social disequilibrium as factors in the explosions. The cities and periods studied are: Florence (1400–1500), Vienna (1880–1930), and New York (1950–1980). See esp. p. 36.

141. Hohenberg and Lees, *The Making of Urban Europe*, p. 207.

142. North, *Institutions, Institutional Change, and Economic Performance*, p. 121.

143. Oliver E. Williamson, "Chester Barnard and the Incipient Science of Organization," in Williamson, *Organization Theory*, pp. 190–99.

144. This point follows from the asset-specificity version of transaction cost theory, but it is not one which Williamson himself emphasizes. He follows Barnard in his conception of the employees of a firm being there by consensus (at least within a certain "zone of indifference" within which they do not mind obeying commands). It is Douglas North who mentions the decreased bargaining power of deskilled workers as a decreased transaction cost for managers, in North, *Institutions, Institutional Change, and Economic Performance*, p. 65.

145. Michael Dietrich, *Transaction Cost Economics and Beyond* (London: Routledge, 1994), pp. 20–28. See also his analysis of the evolution of production methods from the putting-out system to the factory, in terms of his modified transaction cost theory, in ch. 4.

146. Galbraith, *The New Industrial State*, chs. 7 and 15. The classical study of the modern corporation, and of the question of the separation of ownership from control, is Adolf A. Berle and Gardiner C. Means, *The Modern Corporation and Private Property* (New Brunswick, NJ: Transaction, 1991).

147. North traces the origin of this organizational form to the Commenda, of Jewish, Byzantine, and Muslim origins. North, *Institutions, Institutional Change, and Economic Performance*, p. 127. On the Companies of Indias as states within the state, see for example Braudel, *The Perspective of the World*, p. 213.

148. William Lazonick, *Business Organization and the Myth of the Market Economy*

(New York: Cambridge University Press, 1994), p. 5. Like Michael Dietrich (see note 145 above), Lazonick is critical of Williamson's version of transaction cost theory, and offers an expanded version. See also his analysis of why Marxist economic historians failed for a long time to understand this particular organizational form (the joint-stock company with its separation of control from ownership), in ch. 8.

149. Braudel, *The Perspective of the World*, pp. 128–31.

150. Roy Lubove, "Urban Planning and Development," in Kranzberg and Pursell, *Technology in Western Civilization*, vol. 2, p. 462.

151. *Ibid.*, p. 465.

152. *Ibid.*, p. 466.

153. Jean-Francois Hennart, "The Transaction Cost Theory of the Multinational Enterprise," in *The Nature of the Transnational Firm*, eds. Christos N. Pitelis and Roger Sugden (London: Routledge, 1991), p. 85.

154. Herman E. Krooss and Charles Gilbert, *American Business History* (Englewood Cliffs, NJ: Prentice-Hall, 1972), p. 149.

155. *Ibid.*, p. 155.

156. On this form of internalization, see Hennart, "The Transaction Cost Theory of the Multinational Enterprise," pp. 93–95. Internalization of market transactions was practiced by early international firms. Transnational corporations before World War I, whether based in London, Amsterdam, Paris, or Berlin, maintained a small head office in those cities while keeping all their productive assets abroad. These firms were in the business of exporting money, an operation that can be performed in a decentralized way by bank loans and corporate bonds. However, the transaction costs incurred here (screening borrowers for reputation or credit history, demanding collaterals, enforcing payments) may be bypassed by internalizing the borrowing firm. This also increased the power of transnational firms, since by simply lending money they had no control over how the capital would be spent.

157. Harold I. Sharlin, "Electrical Generation and Transmission," in Kranzberg and Pursell, *Technology in Western Civilization*, vol. 1, p. 584.

158. Hennart, "The Transaction Cost Theory of the Multinational Enterprise," pp. 87–88.

159. Peter F. Drucker, "Technological Trends in the Twentieth Century," in Kranzberg and Pursell, *Technology in Western Civilization*, vol. 2, pp. 14–15.

160. This point applies regardless of whether electrical power was generated using the falling water of Niagara Falls or steam turbines:

> Almost at once the turbine began to demonstrate the outstanding economic characteristic of electrical power generation and transmission, the reduction of unit costs with larger size.... It was the greater economy of the larger turbines that eroded the original cost advantage to the manufacturer to generate his own electricity. Along with the opportunities for greater economic efficiency through larger size were those for

greater physical efficiency through higher steam pressures and temperatures, as established in the laws of thermodynamics.... Unit and station size, temperature and pressure all increased with the accumulation of experience, the development of improved materials and techniques, and the growth in power consumption within the separate power systems. (Bruce C. Netschert, "Developing the Energy Inheritance," in Kranzberg and Pursell, *Technology in Western Civilization*, vol. 2, p. 248)

161. A cotton mill in the United States was the first to be completely electrified in 1894, when a central electric motor replaced its central steam motor; this simple substitution, however, was not in itself enough for the new energy form to take over. See Harold I. Sharlin, "Applications of Electricity," in Kranzberg and Pursell, *Technology in Western Civilization*, vol. 1, p. 578.

162. J. A. Duffie, "Energy Resources for the Future," in Kranzberg and Pursell, *Technology in Western Civilization*, vol. 2, p. 288.

163. Hohenberg and Lees, *The Making of Urban Europe*, p. 316.

164. Lubove, "Urban Planning and Development," pp. 474–75.

165. Sharlin, "Electrical Generation and Transmission," p. 585. Sharlin there writes: "Financial backing for the Niagara project came largely from American sources, though most of the nineteenth century American enterprises had been largely dependent on foreign capital, for the most part British. By 1890 American capital was well on its way to independence from foreign sources."

166. Braudel, *The Perspective of the World*, p. 629.

167. Jacobs, *Cities and the Wealth of Nations*, pp. 183–98. There are other "transactions of decline" behind the killing of cities, in which both government and antimarket hierarchies are involved: the war industry. The great intensification represented by wars, at least by the kind of total mobilization of a country's resources which began with Napoleonic warfare, has been widely recognized as a trigger for technological development. Wars, of course, are also a major form of destruction and depletion of resources, which is why the nations that benefit from the greatly intensified flows of matter, energy, and information are those away from the front, like the United States and Japan after World War I. On this point, see Kennedy, *The Rise and Fall of the Great Powers*, p. 279.

However, when military buildups do not occur as short, turbulent spasms, but as a prolonged process during peacetime, they interfere in several ways with economies of agglomeration. For example, they redirect the flow of goods from small towns into garrison cities, like Jacksonville, North Carolina. Jacobs argues that despite the fact that the post exchanges that retail these goods in military towns are the third largest merchandising enterprise in the world, the flow of goods they mobilize is basically sterile (i.e., not part of any autocatalytic dynamics), and its consumption is financed by taxing wealth-producing cities. Thus the economies of agglomeration of big, heterogeneous urban centers are milked by national governments to finance homogenized army towns, while small cities are excluded from the flow of potentially replaceable imports they need to generate their own agglomeration economies. See Jacobs, *Cities and the Wealth of Nations*, pp. 184–87.

168. Drucker, "Technological Trends in the Twentieth Century," p. 11.

169. Gilbert Ryle, *The Concept of Mind* (Chicago: University of Chicago Press, 1984), pp. 27–32. Here Ryle distinguishes between two forms of knowledge, which he calls "knowing that" and "knowing how." With the possible exception of Jean Piaget, the study of skill and other forms of embodied knowledge has been neglected by scientists as well as philosophers. A few studies were conducted in the 1920s, and during World War II, when a great need arose for soldier training techniques, more work was done in the 1940s. Yet the field remained fragmented until the 1970s. See H. T. A. Whiting, *Concepts in Skill Learning* (London: Lepus, 1975), intro. and pp. 3–6.

Economists are finally catching up with know-how and replacing homogeneous rationality with heterogeneous problem-solving skills. See for example, Richard Nelson and Sidney Winter, *An Evolutionary Theory of Economic Change* (Cambridge, MA: Belknap, 1982), pp. 88–90.

170. Galbraith, *The New Industrial State*, pp. 66–67.

171. Annalee Saxenian, "Lessons from Silicon Valley," in *Technology Review* 97.5 (July 1994), p. 44.

172. *Ibid.*, p. 47.

173. Humberto Maturana, "Everything Is Said by an Observer," in *Gaia, a Way of Knowing: Political Implications of the New Biology*, ed. William Irwin Thompson (Hudson, NY: Lindisfarne, 1987), p. 73.

174. A good example of this uncritical attitude toward so-called scientific management is Peter F. Drucker, "Technology and Society in the Twentieth Century," in Kranzberg and Pursell, *Technology in Western Civilization*, vol. 2, p. 25. As Drucker observes, routinization did create economies of scale, which resulted in both lower costs and cheaper prices for products, as well as in higher wages for unskilled jobs (so that both consumers and de-skilled workers benefited somewhat). What he does not consider (or rather, does not value) is the loss of control of the process by the worker and the further de-skilling that went with it (see *ibid.*, p. 26). But as Foucault reminds us, a full cost-benefit accounting of routinized, disciplinary operations needs to be performed not only in terms of economic utility but also in terms of political obedience. And the gains (in terms of economies of scale) may be offset by the costs (in terms of loss of control and de-skilling). I elaborate on this in Chapter Two, but at this point what matters is to emphasize that the "progressive" character of scientific management techniques seemed self-evident not only to people like Drucker but even to those who claimed to be the champions of the working class. It took Marxists almost a century to realize that Taylorism meant the militarization (*not* the "scientifization") of the production process. Lenin, for example,

welcomed scientific management into revolutionary Russia, as one of the "good things" that "capitalism" had created. See Vladimir Lenin, *The Immediate Tests of the Soviet Government,* in *Collected Works,* vol. 27 (Moscow, 1965).

175. On the history of automation, see James R. Bright, "The Development of Automation," in Kranzberg and Pursell, *Technology in Western Civilization,* vol. 2. The evolution of the components of the automated factory took place in the last two centuries, along with the intensification of the process of routinization, and, as with the latter, it involved a constant interplay between military and industrial hierarchies. Bright distinguishes three different components of automation: machines that perform the production operations, machines that move materials from machine to machine in a continuous flow, and a system that controls and coordinates flows and machines. Each of these three componenents evolved more or less independently, finally coming together in the 1940s and 1950s in the United States. The first component, machines that perform operations like cutting, rolling, or mixing, is perhaps the oldest. Bright writes:

> Automatic machines for production actions can be traced back at least to the early 1800's in many fields, and were commonplace in almost every field of manufacturing by the 1870's. In textiles, for example, the industry's history beginning in the early 1700's, reflected mechanization, the application of power to integration of successive operations, and automatic control.... Perhaps the earliest system of automatic machines ... for parts manufacture, as distinct from bulk materials, was the pulley-block machinery built by Marc Brunel for the British Admirality [1802–1808]. (p. 642)

The second component, the automatic handling of a continuous flow of materials at different stages of production, is also very old. Although without a mechanical conveyor belt, some parts of the arsenal of Venice as early as the fifteenth century had a primitive continuous-movement production line. An industrial process using automatic handling with a powered conveyor line "was first identified in the biscuit-baking ... process at the Deptford Victualling Department of the English Navy, 1804–33.... From the 1830's on there were many attempts at continuous process-ing. Feeding hides, sewn together as a continuous sheet, through tanning baths, continuous brickmaking and sugar-cane processing are examples. Processing while moving gradually became a recognized industrial principle contributing to automatic operation" (*ibid.,* p. 647).

Finally, the control of this continuous flow, and its synchronization with the machines that operate on it, evolved from devices like the cam, which forces machines to perform a fixed series of operations. Sophisticated versions of devices like this were used in the 1820s in some American arsenals to control the production of weapon parts. The use of punched cards, as in the loom developed by Jaquard in 1804, allowed the storage of these fixed routines. By World War I, a variety of electric, hydraulic, and pneumatic devices had been created to perform sophisticated control operations, although still in rigid sequences. Adding flexibility to this controlling machinery could be achieved either by using negative feedback — although servomechanisms were not really common outside chemical and electrical plants — or by using programmable computers, but this would have to wait a few decades more (*ibid.*, p. 645).

Routinization is precisely the process through which these three series of operations (to process, to move, to control) were derived in the first place. Human beings, through the daily exercise of their skills, are the source of these operations. But while early-nineteenth-century skilled workers created and controlled their own operations, their counterparts a hundred years later would be executing a fixed, routinized series of actions that someone else had devised for them. In this sense, they were no different from the machines that would soon replace them. In a slowly intensifying process of routinization, first the production, later the control operators were taken out of the autocatalytic loop and reduced to its external triggers, a set of completely de-skilled button pushers. The development of "scientific management" by Frederick Taylor at the turn of the century, in which a worker's operations were carefully broken down into their components and put back together again into a series of optimized, homogenized operations, represents the peak of this intensification. See, for example, Braverman, *Labor and Monopoly Capital*, ch. 8.

176. See, for example, Thomas W. Malone and John F. Rockart, "Computers, Networks and the Corporation," in *Scientific American* 265.3 (September 1991), p. 131. One of the examples of agglomeration economies mentioned in this article, a series of textile firms near Prato, Italy, studied by Michael Piore and Charles Sabel, is also mentioned (as alternative to antimarkets and economies of scale) in Braudel, *The Perspective of the World*, p. 630, and Jacobs, *Cities and the Wealth of Nations*, p. 40.

177. Jacobs, *Cities and the Wealth of Nations*, pp. 45–49.

178. Richard J. Barnett and Ronald E. Muller, *Global Reach: The Power of the Multinational Corporations* (New York: Simon and Schuster, 1974), p. 40. However, the authors assume that all this is explainable in terms of the "laws of capitalism" and neglect to mention the role of military institutions in the development of operations research during World War II. On that point, see Stephen P. Waring, *Taylorism Transformed: Scientific Management Theory since 1945* (Chapel Hill: University of North Carolina Press, 1991), ch. 2.

CHAPTER TWO: FLESH AND GENES

1. Ian. G. Simmons, *Biogeography: Natural and Cultural* (London: Edward Arnold, 1979), p. 79.

2. *Ibid.*, pp. 70–72.

3. Paul Colinvaux, *Why Big Fierce Animals Are Rare* (Princeton, NJ: Princeton University Press, 1978), pp. 26–27.

4. Simmons, *Biogeography*, p. 67.

5. James H. Brown, "Complex Ecological Systems," in *Complexity: Metaphors, Models and Reality*, eds. George Cowan, David Pines, and David Meltzer (Reading, MA: Addison-Wesley, 1994), p. 424.

6. C. S. Holling, "Resilience and Stability of Ecosystems," in *Evolution and Consciousness*, eds. Erich Jantsch and Conrad Waddington (New York: Addison-Wesley, 1976), pp. 81–82.

7. On cities as "heat islands," see Joseph M. Moran and Michael D. Morgan, *Meteorology* (New York: Macmillan, 1986), pp. 274–76.

8. Thomas F. Glick, "Science, Technology and the Urban Environment," in *Historical Ecology*, ed. Lester J. Bilsky (New York: Kennikat, 1980), p. 126.

9. Fernand Braudel, *Capitalism and Material Life* (New York: Harper and Row, 1973), p. 376.

10. *Ibid.*, p. 377.

11. Simmons, *Biogeography*, pp. 192–93.

12. *Ibid.*, pp. 196–97. See also Alfred W. Crosby, *Ecological Imperialism: The Biological Expansion of Europe, 900–1900* (New York: Cambridge University Press, 1989), pp. 173–74.

13. Braudel, *Capitalism and Material Life*, p. 34.

14. William H. McNeill, *Plagues and Peoples* (Garden City, NJ: Anchor/Doubleday, 1976), p. 45.

15. Claude Lévi-Strauss, *The Raw and the Cooked* (Chicago: University of Chicago Press, 1983).

16. Braudel, *Capitalism and Material Life*, p. 39.

17. Richard Dawkins, *The Selfish Gene* (New York: Oxford University Press, 1990), pp. 19–20.

18. James D. Watson, *Molecular Biology of the Gene* (Menlo Park, CA: W.A. Bemjamin, 1970), p. 145. Here Watson observes: "Enzymes never affect the nature of an equilibrium: They merely speed up the rate at which it is reached. Thus, if the thermodynamic equilibrium is unfavourable for the formation of a molecule, the presence of an enzyme can in no way bring about its accumulation."

The dependence catalysts (and hence genes) exhibit with respect to energy flows becomes even more pronounced when the thermodynamics involved are far from equilibrium. That is, in these conditions genes (or their phenotypic effects) become mere switching devices to pick one among multiple coexisting equilibriums. Moreover, the catalysts themselves are subject to a nonlinear combinatorics; that is, they may enter into self-sustaining autocatlytic loops, with their own internal coherence. All this is particularly clear in the case of the embryological process: the transformation of a single-cell egg into a complex multicellular organism.

Basically, at the beginning of the transformation the egg may be seen as an enclosed portion of the flows of genes and biomass, that is, as a nucleus and a cytoplasm. The latter is a source of food, as well as a complex nonlinear dynamical system with multiple equilibriums. It is this energetic flesh that is the seat of processes of self-organization. For example, if the genetic information in the nucleus is removed from a fertilized egg (or neutralized), the cell still undergoes some of its bifurcations between stable states (i.e., gastrulation). See Vladimir Glisin, "Molecular Biology in Embryology: The Sea Urchin Embryo," in *Self-Organizing Systems*, ed. Eugene Yates (New York: Plenum, 1987), p. 163.

The remaining stable states, the final form cells from different tissues take (e.g,. bone, muscle, or nerve cells), may also be nonlinear stable states, this time of the dynamics of meshworks of gene products (enzymes) or meshworks of regulator genes. See Stuart Kauffman, *The Origins of Order: Self-Organization and Selection in Evolution* (New York: Oxford University Press, 1993), p. 525.

At higher levels — tissues, organs, organisms — attractors are also postulated. Here it is "morphogenetic fields" that do the attracting. (The concept of this kind of field derives, in fact, from very early interactions between nonlinear mathematics [Rene Thoms's catastrophe theory] and embryology [Waddington].) The thrust of this early current of nonlinear biology is now provided by people such as Brian Goodwin. See Brian Goodwin, "Developing Organisms as Self-Organizing Fields," in Yates, *Self-Organizing Systems*, p. 176.

19. Howard Pattee, "The Problem of Biological Hierarchy," in *Towards a Theoretical Biology*, ed. C. H. Waddington (Edinburgh: Edinburgh University Press, 1968).

20. As the philosopher of science Elliot Sober puts it, natural animal populations are intrinsically variable: "Uniformity . . . takes some work. Natural Selection is one mechanism that can *destroy* variation. For it to act at all there must be variation (in fitness). But once a selection process begins, it gradually destroys the conditions needed for its continuing operation. Selection eliminates variation in fitness, and thereby brings itself to a halt" (Elliot Sober, *The Nature of Selection: Evolutionary Theory in Philosophical Focus* [Cambridge, MA: MIT Press, 1987], p. 159).

21. William H. Durham, *Coevolution: Genes, Culture, and Human Diversity* (Stanford, CA: Stanford University Press, 1991), chs. 3 and 6.

22. Richard Lewontin, *Human Diversity* (Scientific American Books, 1982), p. 123.

23. *Ibid.*, pp. 115–17.

24. D. F. Roberts, "Migration in the Recent Past: Societies with Records," in *Biological Aspects of Human Migration*, eds. C. G. N. Mascie-Taylor and G. W. Lasker (Cambridge, MA: Cambridge University Press, 1988), p. 67.

25. Kenneth M. Weiss, "In Search of Times Past: Gene Flow and Invasion in the Generation of Human Diversity," in *ibid.*, p. 148.

26. Luigi Cavalli Sforza, "Diffusion of Culture and Genes," in *Issues in Biological Anthropology*, ed. B. J. Williams (Malibu, CA: Undena, 1986), pp. 13–14. On the general issue of the competition between "diffusion of ideas" and "migration of bodies and culture" as explanatory paradigms in anthropology, see William Y. Adams, "On

Migration and Diffusion as Rival Paradigms," in *Diffusion and Migration: Their Roles in Cultural Development*, eds. P. G. Duke, J. Ebert, G. Langeman, and A. P. Buchner (Calgary: University of Calgary, 1978), pp. 1–5. These questions are related to the issue of "cultural relativism," which I criticize below (particularly in its modern cliché version: "everything is socially constructed"). The same anthropologists who wrongly banished all biological issues from consideration also promoted "diffusionism" as the only valid explanation. See below, note 96.

27. Weiss, "In Search of Times Past," p. 149.

28. Roberts, "Migration in the Recent Past," p. 62.

29. Lewontin, *Human Diversity*, p. 113.

30. Barry Bogin, "Rural-to-Urban Migration," in Mascie-Taylor and Lasker, *Biological Aspects of Human Migration*, p. 93.

31. Paul M. Hohenberg and Lynn Hollen Lees, *The Making of Urban Europe, 1000–1950* (Cambridge, MA: Harvard University Press, 1985), p. 89.

32. Paul Colinvaux, *The Fates of Nations: A Biological Theory of History* (New York: Simon and Schuster, 1980), p. 70. I adopt here a few of Colinvaux's views (e.g., his theory of social niches) but by no means his entire outlook, which is too deterministic. He attempts to reduce the diversity of forces operating in human history to a few ecological determinants, and this impoverishes his theory.

33. *Ibid.*, pp. 39–44.

34. Hohenberg and Lees, *The Making of Urban Europe*, pp. 86 and 97.

35. *Ibid.*, pp. 79–80.

36. Bryant Robey, Shea O. Rustein, and Leo Morris, "The Fertility Decline in Developing Countries," in *Scientific American* 269.6 (December 1993), p. 60.

37. Michele Wilson and Frances A. Boudreau, "The Sociological Perspective," in *Sex Roles and Social Patterns*, eds. Frances A. Boudreau, Roger S. Sennott, and Michele Wilson (New York: Praeger, 1986), p. 8.

38. Marvin Harris, *Cannibals and Kings* (New York: Vintage, 1991), ch. 6. I do not mean to imply that this is the only, or best, theory of the origin of reproductive strata. It is, however, the one most relevant to my subject here (the urban Middle Ages), since it is precisely the exclusion from warrior roles that is behind the function of guardianship.

39. Lewontin, *Human Diversity*, p. 109.

40. Edith Ennen, *The Medieval Woman* (Oxford, UK: Basil Blackwell, 1989), p. 267.

41. *Ibid.*, p. 36.

42. *Ibid.*, p. 279.

43. *Ibid.*, p. 101.

44. Braudel, *Capitalism and Material Life*, p. 410. Not only cities affected class structure; the latter often influenced urbanization as well. As Braudel points out:

The social structures of both India and China automatically rejected the town and offered, as it were, refractory, sub-standard material for it. Therefore if the town did not win its independence it was not only because of the mandarins' beatings or the prince's cruelty to merchants and ordinary citizens. It was because society was well and truly frozen in a sort of irreducible system, a previous crystallization. In the Indies the caste system automatically divided and broke up every urban community. In China the cult of the gentes was opposed to a mixture comparable to that which created the Western town — a veritable machine for breaking up old bonds. (p. 410)

45. *Ibid.*, p. 403.
46. Charles R. Bowlus, "Ecological Crisis in Fourteenth-Century Europe," in Bilsky, *Historical Ecology*, p. 94.
47. *Ibid.*, p. 96.
48. *Ibid.*, p. 89.
49. Vernon Hill Carter and Tom Dale, *Top Soil and Civilization* (Norman: University of Oklahoma Press, 1974), pp. 7–8.
50. *Ibid.*, pp. 138–45; and J. Donald Hughes, *Ecology in Ancient Civilizations* (Albuquerque: University of New Mexico Press, 1975), pp. 116–17.
51. Bowlus, "Ecological Crisis in Fourteenth-Century Europe," p. 96.
52. Braudel, *Capitalism and Material Life*, p. 19.
53. McNeill, *Plagues and Peoples*, p. 103.
54. *Ibid.*, p. 97. Here McNeill writes:

We may infer that by about the beginning of the Christian era, at least four divergent civilized disease pools had come into existence, each sustaining infections that could be lethal if let loose among populations lacking any kind of prior exposure or accumulated immunity. All that was needed to provoke a spillover from one pool to another was some accident of communication permitting a chain of infection to extend to new ground where populations were also sufficiently dense to sustain the infection either permanently, or at least for a season or two.... When ... travel across the breadth of the Old World from China to India to the Mediterranean became regularly organized on a routine basis ... the possibility of homogenization of those infections ... opened up. It is my contention that something approximating this condition did in fact occur, beginning in the first century A.D.

55. *Ibid.*, p. 116.
56. *Ibid.*, p. 146.
57. *Ibid.*, p. 150.
58. Braudel, *Capitalism and Material Life*, pp. 48–49.
59. *Ibid.*, p. 48
60. McNeill, *Plagues and Peoples*, pp. 163–64.

61. *Ibid.*, p. 152. There McNeill writes:

After the Great Plague of London, in 1665, *Pasteurella pestis* withdrew from northwestern Europe.... Quarantine and other public health measures probably had less decisive overall effect in limiting the outbreaks of plague, whether before or after 1665, than other unintended changes in the manner in which European populations coexisted with fleas and rodents. For instance, in much of western Europe, wood shortages led to stone and brick house construction, and this tended to increase the distance between rodent and human occupants of the dwelling, making it far more difficult for a flea to transfer from a dying rat to a susceptible human. Thatch roofs, in particular, offered ready refuge for rats; and it was easy for a flea to fall from such a roof onto someone beneath. When thatch roofs were replaced by tiles... opportunities for this kind of transfer of infection drastically diminished.

62. Braudel, *Capitalism and Material Life*, p. 38.

63. Fernand Braudel, *The Perspective of the World* (New York: Harper and Row, 1986), p. 117.

64. Archibald Lewis, "Ecology and the Sea in the Medieval Times (300–1500 A.D.)," in Bilsky, *Historical Ecology*, p. 74.

65. Ian G. Simmons, *Changing the Face of the Earth: Culture, Environment, History* (Oxford, UK: Basil Blackwell, 1989), p. 166.

66. Braudel, *Capitalism and Material Life*, p. 268.

67. Braudel, *The Perspective of the World*, p. 108.

68. *Ibid.*, p. 157.

69. Carter and Dale, *Top Soil and Civilization*, pp. 151 and 174.

70. Braudel, *The Perspective of the World*, p. 89.

71. *Ibid.*, p. 177.

72. Elio Conti, mentioned in Fernand Braudel, *The Wheels of Commerce* (New York: Harper and Row, 1983), p. 256.

73. Braudel, *Capitalism and Material Life*, p. 373.

74. Braudel, *The Wheels of Commerce*, pp. 229–30.

75. McNeill, *Plagues and Peoples*, p. 6. On the theme of micro- and macroparasites, see also William H. McNeill, *The Human Condition: An Ecological and Historical View* (Princeton, NJ: Princeton University Press, 1980).

76. Braudel, *The Wheels of Commerce*, pp. 265–72.

77. Crosby, *Ecological Imperialism*, p. 63.

78. *Ibid.*, p. 65.

79. McNeill, *Plagues and Peoples*, p. 62.

80. *Ibid.*, p. 63.

81. Crosby, *Ecological Imperialism*, p. 99.

82. *Ibid.*, p. 52.

83. McNeill, *Plagues and Peoples*, p. 178.

84. *Ibid.*, p. 180. See also Pierre Chaunu, quoted in Pierre Clastres, *Society against the State* (New York: Zone, 1987), p. 99. There Clastres quotes Chaunu's claim: "It appears that one-fourth of mankind was annihilated by the microbic shocks of the sixteenth century."

85. McNeill, *Plagues and Peoples*, p. 185.

86. Braudel, *Capitalism and Material Life*, p. 344.

87. Niles Eldridge, *Macroevolutionary Dynamics: Species, Niches, and Adaptive Peaks* (New York: McGraw-Hill, 1989), pp. 104–105; and James L. Gould and Carol G. Gould, *Sexual Selection* (New York: Scientific American Library, 1989), pp. 80–105.

88. On the role of sexual selection, see Richard Dawkins, *The Selfish Gene*, p. 158.

89. On the role of retroviruses in evolution, see E. J. Steele, *Somatic Selection and Adaptive Evolution* (Chicago: University of Chicago Press, 1981), pp. 47–50. Dawkins accepts the existence of these horizontal gene transmissions but rejects the idea that they imply Lamarkism (inheritance of acquired traits) as opposed to a kind of "somatic Darwinism." See Richard Dawkins, *The Extended Phenotype* (Oxford, UK: Oxford University Press, 1990), pp. 166–72. Gilles Deleuze and Félix Guattari also mention this phenomenon, which to them provides evidence that the evolutionary "tree" is more like a rhizome. See Gilles Deleuze and Félix Guattari, *A Thousand Plateaus* (Minneapolis: University of Minnesota Press, 1987), p. 10.

90. K. W. Jeon and J. F. Danielli, quoted in Richard Dawkins, *The Extended Phenotype*, pp. 159–60.

91. John H. Holland, *Adaptation in Natural and Artificial Systems* (Cambridge, MA: MIT Press, 1992), chs. 9 and 10.

92. The term "meme" was introduced in Dawkins, *The Selfish Gene*, ch. 11. However, the concept needs further elaboration not only to distinguish it from other replicators (such as linguistic norms) but even for application to animal protocultures, since it is hard to show that "true" imitation occurs in the wild. See Kevin N. Laland, Peter J. Richardson, and Robert Boyd, "Animal Social Learning: Towards a New Theoretical Approach," in *Perspectives in Ethology*, eds. P. P. G. Bateson, Peter H. Klopfer, and Nicholoas S. Thompson (New York: Plenum, 1993). On the use of memes to investigate animal protocultures, see John T. Bonner, *The Evolution of Culture in Animals* (Princeton, NJ: Princeton University Press, 1980), ch. 2.

93. Dawkins, *The Selfish Gene*, p. 24. Here Dawkins observes: "Genes have no foresight. They do not plan ahead. Genes just *are*, some genes more so than others, and that is all there is to it." That is, genes are just *replicators*, and some replicate more than others.

94. Philosophically, besides showing that one and the same abstract machine is behind many different types of phenomena and that therefore it is not what gives a given phenomenon its identity (i.e., it does not constitute the essence of *that* phenomenon), we also need to show that the relation between an abstract machine

and the concrete assemblages that instantiate it is not one of "transcendence" but one of "immanence." In other words, we need to show that abstract machines do not exist in some transcendental heaven waiting to be incarnated in concrete mechanisms, but that they are *intrinsic* features of matter-energy flows subject to nonlinear dynamics and nonlinear combinatorics. This is, I believe, the position adopted by Deleuze and Guattari. See, for example, Deleuze and Guattari, *A Thousand Plateaus*, pp. 266–67.

The simplest examples of abstract machines, such as periodic attractors (capable of very diverse instantiations: crystal radios, chemical clocks, Kondratiev waves in the economy, etc.) are the easier to explain in these nontranscendental terms. See, for example, Gregoire Nicolis and Ilya Prigogine, *Exploring Complexity* (New York: W.H. Freeman, 1989), p. 100.

95. Stuart Kauffman, *The Origins of Order: Self-Organization and Selection in Evolution* (New York: Oxford University Press, 1993), chs. 3 and 6.

96. In the 1980s many of the original "discoveries" of cultural anthropologists were found to be oversimplifications or even distortions of the social realities they had studied. (The most famous debunkings were perhaps of Margaret Mead's claims that adolescents in Samoa did not go through similar anxieties as their Western counterparts and that males and females in Chambri exhibited an opposite pattern of dominance as in most other societies.) On all this, and the process through which cultural relativism became entrenched in academic circles, see Donald E. Brown, *Human Universals* (New York: McGraw-Hill, 1991).

The "debunker," in the case of Mead's observations on Samoa, was the anthropologist Derek Freeman. However, this cannot be boiled down to a question of different interpretations of the data, each having an equal chance of being valid. As Brown puts it, "Mead's book was based on 9 months of fieldwork when she was 23 years old. Derek Freeman . . . conducted 6 years of fieldwork in Samoa" (*ibid.*, p. 16). He then adds: "One can only ask how Mead could have been so wrong. . . . Mead went to Samoa without a knowledge of the language and with unfortunate gaps in her familiarity with the extensive literature on Samoa. . . . When she reached Samoa she did not undertake a general study of the Samoan ethos and culture but launched directly into her study of adolescence. Her informants were adolescent girls; neither boys nor adults were studied" (*ibid.*, pp. 18–19).

The list of criticisms continues. One can only wonder how the modern left (or rather, that influential segment of it, the "social constructionists") can pretend to offer a coherent strategy of resistance based on such flimsy foundations. In any event, the fortress walls of cultural relativism will prove a poor defense against the new dangers posed by human sociobiologists. Indeed, the old stance may actually be counterproductive since it will make any revelation of its inadequacies (as in the case of the universality of color perception or facial expressions) seem like a triumph for the opposition.

Although neither one of these two universals makes a difference in terms of the stance taken in this book, there is another one that does. In Chapter One I adopted the revised form of "methodological individualism" created by neoinstitutionalist economists. Unlike the neoclassical economics version (which is "atomistic"), the new version is compatible with "ontological holism." Hence, it rejects the view of isolated individuals as decision makers and brings into the picture collective entities such as institutional rules guiding decision making. What does survive from the old view, however, is the idea that human individuals are basically "self-interested." This is, of course, a basic theme in human and animal sociobiology. (Dawkins, in his book on "selfish genes," demolishes the claims of group selectionism, which assumes naturally occurring "altruistic tendencies" in animals.) Self-interest, however, should in no way be construed as an "essence" but rather as a multitrack disposition with evolutionary-historical (and hence contingent) origins. To put it bluntly, if humans reproduced by cloning (or even, as ants or bees do, through a special caste of reproducers) this selfishness would not exist as it does. And, at any rate, the point here is not that we must accept this as a biological destiny, but that we must face the responsibility that this organic constraint imposes on us: children do not come equipped with "altruism" and must be taught to share and to respect others. Finally, self-interest is perfectly compatible, in both animals and humans, with reciprocating behavior. (And, clearly, cultural constraints may overwhelm the biological ones, in special circumstances.) See also my remarks in note 103 below.

97. Durham, *Coevolution*, p. 187.

98. Stephen Jay Gould, *The Mismeasure of Man* (New York: W.W. Norton, 1981), p. 324. This book stands out as a perfect example of how criticism of science (in this case, scientific racism) should be made. This valid line of criticism must be opposed to the claim that a critique of science can be carried out with empty assertions such as "science is socially constructed," which is either redundant (everything produced by human beings is a social product) or false (if taken to imply that the epistemological status of scientific statements is the same as that of other cultural products, such as religion).

99. Durham, *Coevolution*, pp. 213–23.

100. *Ibid.*, p. 283.

101. *Ibid.*, p. 289.

102. *Ibid.*, p. 362.

103. *Ibid.*, p. 164, and Brown, *Human Universals*, p. 66. The question of "inclusive fitness" serves as a good example to illustrate this point. The term is used in animal (and human) sociobiology to explain the simple Darwinian logic behind the altruistic behavior of parents toward their offspring. Basically, the idea is that much as genes connected to behavior that increases the reproductive success of their carriers will tend to accumulate in a population, so will genes that promote the

reproductive success of one's offspring. If we view genes simply as material replicators, then, simply in terms of intensity of flow, half of the genes that parents
passed to their children will also be passed to their grandchildren, so whatever
sorting processes benefit the sedimentation of the former will also favor (propor-
tionally) the latter. However, a cultural relativist could dismiss this logic by arguing,
quite correctly, that kinship structures (that is, the social rules determining what
counts as a "close relative") vary across cultures and that they determine kinship
relations in ways that not only do not correspond to the expectations of the inclusive fitness hypothesis but even work against it sometimes. However, as the anthropologist Napoleon Chagnon puts it, "the structuralist approach to kinship tends to
view the system as an 'ideal' or 'perfect' system of classification and it is not concerned with individual conformity or deviance.... Most anthropologists are aware,
however, that there are always some discrepancies between 'rules' and 'behavior'
in all realms of culture.... Manipulating or fudging of genealogies, for example, is
commonly reported in the ethnographic literature." See Napoleon Chagnon, "Male
Yanomamo Manipulations of Kinship Classifications of Female Kin for Reproductive
Advantage," in *Human Reproductive Behaviour: A Darwinian Perspective*, eds. Laura
Betzig, Monique Borgerhoff Mulder, and Paul Turke (Cambridge, UK: Cambridge
University Press, 1988), p. 25. Chagnon shows how, when we take into account this
manipulation of the cultural rules that define kinship, the actual behavior of human
beings is closer to what inclusive fitness would predict (or, rather, the result of an
interaction between cultural and organic constraints).

104. Sforza, "Diffusion of Culture and Genes," p. 30.

105. *Ibid.*, pp. 31–32.

106. Braudel, *Capitalism and Material Life*, pp. 294–98.

107. Richard Nelson and Sidney Winter, *An Evolutionary Theory of Economic
Change* (Cambridge, MA: Belknap, 1982), pp. 98–100.

108. Crosby, *Ecological Imperialism*, p. 300.

109. *Ibid.*, p. 12.

110. *Ibid.*, pp. 148–49.

111. On the stock effect, see William J. Smyth, "Irish Emigration, 1700–1920,"
in *European Expansion and Migration*, eds. P. C. Emmer and M. Morner (Oxford, UK:
Berg, 1992), p. 58. For a criticism of too mechanical a view of push (famines) and
pull factors (stock effect), see Magnus Morner, "Divergent Perspectives," in *ibid.*,
p. 277.

112. Crosby, *Ecological Imperialism*, pp. 288–89.

113. *Ibid.*, pp. 290–91.

114. *Ibid.*, p. 170.

115. *Ibid.*, p. 151.

116. *Ibid.*, p. 176.

117. *Ibid.*, pp. 177–79.

118. McNeill, *Plagues and Peoples*, p. 193.

119. P. C. Emmer, "European Expansion and Migration: The European Colonial Past and Intercontinental Migration; An Overview," in Emmer and Morner, *European Expansion and Migration*, pp. 10–12.

120. Crosby, *Ecological Imperialism*, p. 302.

121. Emmer, "European Expansion and Migration," p. 8.

122. Crosby, *Ecological Imperialism*, p. 305.

123. Braudel, *The Perspective of the World*, p. 388.

124. Braudel, *The Wheels of Commerce*, pp. 265–72.

125. Sidney W. Mintz, *Sweetness and Power: The Place of Sugar in Modern History* (New York: Viking, 1985), p. 188.

126. *Ibid.*, p. 191.

127. Braudel, *The Wheels of Commerce*, pp. 277–78.

128. McNeill, *Plagues and Peoples*, p. 223.

129. *Ibid.*, p. 221.

130. *Ibid.*, p. 223.

131. *Ibid.*, p. 210; and Michel Foucault, *Discipline and Punish: The Birth of the Prison* (New York: Vintage, 1979), p. 186.

132. Foucault, *ibid.*, p. 144.

133. *Ibid.*, p. 199.

134. McNeill, *Plagues and Peoples*, p. 155.

135. Foucault, *Discipline and Punish*, p. 198.

136. *Ibid.*, p. 190.

137. McNeill, *Plagues and Peoples*, pp. 234–35.

138. Foucault, *Discipline and Punish*, pp. 138–39. Organization theorists have uncovered several ways in which routines may spread contagiously through an ecology of institutions:

> The first is the diffusion involving a single source broadcasting a disease to a population of potential, but not necessarily equally vulnerable, victims. Organizational examples include rules promulgated by governmental agencies, trade associations, professional associations, and unions. The second process is diffusion involving the spread of a disease through contact between a member of the population who is infected and one who is not, sometimes mediated by a host carrier. Organizational examples include routines diffused by contacts among organizations, by consultants, and by the movement of personnel. The third process is two-stage diffusion involving the spread of a disease within a small group by contagion and then, by broadcast from them to the remainder of the population. Organizational examples include routines communicated through formal and informal educational institutions, experts, and trade and popular publications. In the organizational literature, these three processes have been labeled coercive, mimetic, and normative. (Barbara Levitt and James G. March, "Chester I.

Barnard and the Intelligence of Learning," in *Organization Theory*, ed. Oliver E. Williamson [New York: Oxford University Press, 1995], p. 25)

On the other hand, depending on the particular mix of technological and institutional factors in the form of a given individual organization, the motivation for the incorporation of outside routines may vary, from a strict economic motivation framed in terms of increased efficiency to one framed in terms of increased legitimacy. One author believes that it was precisely this search for legitimacy (of a given organizational form in an organizational ecology) that may account for the uniform transfer of institutional innovations in schools, hospitals, and other institutions. See W. Richard Scott, "Symbols and Organizations: From Barnard to the Institutionalists," in *ibid.*, p. 49.

139. William H. McNeill, *The Pursuit of Power: Technology, Armed Force, and Society since A.D. 1000* (Chicago: University of Chicago Press, 1982), p. 129.

140. Foucault, *Discipline and Punish*, p. 138. This institutional transformation can be further characterized as involving a double inversion of relations of visibility. While the old institutions (the leprosaria or the dungeons) vanished individuals from sight, modern hospitals and prisons made their bodies much more visible and analyzable. On the other hand, while the old forms of power manifested themselves through spectacular displays, such as public executions, the new strategies made the application of power almost invisible. Control over the human body now took the form of drills and exercises, endlessly repeated routines through which conformity and obedience to a norm were elicited. And if punishment was involved, it was not in the highly visible form of torture, but in less obvious and yet more effective ways — light physical punishment and minor deprivations — but used constantly to penalize even slight departures from routines and norms.

141. McNeill, *The Pursuit of Power*, p. 147.

142. Braudel, *The Wheels of Commerce*, pp. 284–86.

143. Clyde Manwell and C. M. Ann Baker, *Molecular Biology and the Origin of Species: Heterosis, Protein Polymorphism and Animal Breeding* (Seattle: University of Washington Press, 1970), p. 315.

144. *Ibid.*, p. 317.

145. G. E. Fussell, "The Agricultural Revolution, 1600–1850," in *Technology in Western Civilization*, 2 vols., eds. Melvin Kranzberg and Carrol W. Pursell (New York: Oxford University Press, 1967), vol. 1, p. 137.

146. Henri Pirenne quoted in Jan De Vries, *The Dutch Rural Economy in the Golden Age, 1500–1700* (New Haven, CT: Yale University Press, 1974), p. 3.

147. McNeill, *Plagues and Peoples*, p. 34.

148. De Vries, *The Dutch Rural Economy in the Golden Age*, p. 149.

149. De Vries offers an argument for the noncapitalistic character of the new system in the case of the northern Netherlands, where a modified version was

developed before it became the core of the English revolution. He develops two simple models to capture the dynamics of farmer response to increased rural population. The first one (which he calls the "peasant model") can be described thus: greater numbers of farmers are accommodated by dividing up the land into smaller plots, each cultivated intensively (careful plowing, weeding, and fertilizing) but still aiming at self-sufficiency (as opposed to connecting with the markets). The labor-intensive character of this strategy, however, meant that farmer productivity actually declined in the transition period, making them more vulnerable to famines and to the macroparasitism of antimarkets and aristocracies, which took advantage of the situation to amass land and revise leasing contracts. A second scenario (which he calls the "specialization model") involves turning to specialized crops aimed at urban markets while the farmers themselves keep control of the process. "The predatory role of capitalists and noblemen in the peasant model has no counterpart in the specialization model since peasants themselves reorganize production in response to market opportunities and themselves reap the benefits" (De Vries, *The Dutch Rural Economy in the Golden Age*, p. 8).

De Vries argues that it is this second model that applies to the Netherlands from the seventeenth century on, even if a large number of other factors (field patterns, legal system, family structure) need to be taken into consideration to account for regional variations. Also, the new farms offered opportunities for investment by Amsterdam's wealthy classes, so subtler forms of antimarket infiltration also took place. Yet, despite their many mixtures in practice, markets and antimarkets must be kept as separate elements in our historical reconstructions.

150. Fussell, "The Agricultural Revolution," p. 142.

151. Georg Borgstrom, "Food and Agriculture in the Nineteenth Century," in Kranzberg and Pursell, *Technology in Western Civilization*, vol. 1, p. 409. There Borgstrom writes:

> By the end of the War of 1812, the land along the eastern seaboard was under cultivation and many new emigrants were settling between the Appalachians and the Mississippi. The drive continued with increasing force until the entire area along the Mississippi River from the Great Lakes to the Gulf was under cultivation. It was the fertile prairies, however, already encountered in Illinois, that encouraged the beginnings of large-scale farming operations which have become characteristic of our own time. This movement was stimulated in the second half of the century by emerging urbanization, swelling immigration, strong population growth, lively industrial expansion, intense railroad construction, and the final settlement of the western frontier.

On the ecological impact of this deforestation, see Carl H. Moneyhon, "Environmental Crisis and American Politics, 1860–1920," in Bilsky, *Historical Ecology*, pp. 141–42.

152. Simmons, *Changing the Face of the Earth*, p. 243.

153. See, for example, Borgstrom, "Food and Agriculture in the Nineteenth Century," p. 413.

154. Simmons, *Biogeography*, p. 231.

155. Jack Doyle, *Altered Harvest: Agriculture, Genetics and the Fate of the World's Food Supply* (New York: Viking, 1985), pp. 34–37.

156. *Ibid.*, p. 42. Hybrid vigor is a nonlinear, emergent property of certain genetic combinations that can occur spontaneously, not just in the controlled laboratory conditions that gave rise to the new corn varieties. It is, indeed, a meshwork effect: it involves an autocatalytic loop of enzymes within the plant or animal body, a loop that has barely crossed the threshold of self-sustainability. In those circumstances the loss of a gene that codes for one of the enzymes can lead to the collapse of the entire loop. However, and for the same reason, mating with an individual which carries that one gene leads to a sudden reassembly of the loop and of the anatomical and physiological traits that depend, directly or indirectly, on the activity and products of the mutually stimulating set of enzymes. See Manwell and Baker, *Molecular Biology and the Origin of Species*, pp. 265–66.

157. Doyle, *Altered Harvest*, p. 2.

158. *Ibid.*, p. 43.

159. Gena Corea, *The Mother Machine: Reproductive Technologies from Artificial Insemination to Artificial Wombs* (New York: Harper and Row, 1986), pp. 17–18.

160. Daniel J. Kevles, "Out of Eugenics: The Historical Politics of the Human Genome," in *The Code of Codes: Scientific and Social Issues in the Human Genome Project*, eds. Daniel J. Kevles and Leroy Hood (Cambridge, MA: Harvard University Press, 1992), p. 6.

161. Gould, *The Mismeasure of Man*, pp. 231–32. On the homogeneity of human gene pools, Gould observes: "the remarkable lack of genetic differentiation among human groups—a major biological basis for debunking determinism—is a contingent fact of evolutionary history, not an a priori or necessary truth. The world might have been ordered differently. Suppose, for example, that one or several species of our ancestral genus *Australopithecus* had survived.... We—that is, *Homo sapiens*—would then have faced all the moral dilemmas involved in treating a human species of distinctly inferior mental capacity" (pp. 322–23).

162. *Ibid.*, p. 155.

163. *Ibid.*, p. 229.

164. Corea, *The Mother Machine*, p. 305.

165. *Ibid.*, p. 315.

166. *Ibid.*, p. 306.

167. McNeill, *Plagues and Peoples*, p. 239.

168. *Ibid.*, p. 240.

169. *Ibid.*, pp. 231–33.

170. J. D. Murray, *Mathematical Biology* (Berlin: Springer Verlag, 1989), p. 657.

171. McNeill, *Plagues and Peoples*, p. 248.

172. Jean Florent and Pierre-Etienne Bost, "The Great Turning Point: Antibiotics and Secondary Metabolites," in *Biotechnology*, eds. Elizabeth Antebi and David Fishlock (Cambridge, MA: MIT Press, 1986), p. 20.

Penicillin, for example, is produced by fermentation, a nutritional strategy more ancient than photosynthesis. Human beings have long used fermentation to create a wide variety of foodstuffs (cheese, yogurt, bread, beer, and wine), and they did this by (unknowingly) recruiting biological catalysts (enzymes) to perform the necessary operations. A piece of the machinery of food webs was literally detached and converted into a source of enzymes, as when goat or sheep stomachs were used to create cheese. Penicillin is not an enzyme but rather a secondary substance (metabolite) produced by a fungus to interfere with the action of the enzymes of other microorganisms. Mass-producing penicillin meant domesticating these fungi, that is, screening candidates from a heterogeneous population (from soil or water samples) and then improving strains by inducing mutations and promoting the propagation of the useful ones.

173. *Ibid.*, p. 22.

174. Simmons, *Changing the Face of the Earth*, p. 262.

175. Elizabeth Antebi and David Fishlock, "The Engineers of Life and Their Chimeras: Recombinant DNA," in Antebi and Fishlock, *Biotechnology*, p. 54.

176. Doyle, *Altered Harvest*, pp. 116–17.

177. *Ibid.*, pp. 261–63.

178. *Ibid.*, p. 138.

179. *Ibid.*, p. 205.

180. *Ibid.*, p. 216.

181. Corea, *The Mother Machine*, pp. 22–23.

182. Dorothy Nelkin and Laurence Tancredi, *Dangerous Diagnostics: The Social Power of Biological Information* (New York: Basic, 1989), p. 176.

CHAPTER THREE: MEMES AND NORMS

1. William Labov, "The Social Setting of Linguistic Change," in William Labov, *Sociolinguistic Patterns* (Philadelphia: University of Pennsylvania Press, 1972), p. 271.

2. M. L. Samuels, *Linguistic Evolution* (London: Cambridge University Press, 1972), p. 90.

3. Martin Harris, "The Romance Languages," in *The Romance Languages*, eds. Martin Harris and Nigel Vincent (New York: Oxford University Press, 1988), p. 5.

4. Alberto Varvaro, "Latin and Romance: Fragmentation or Restructuring?" in *Latin and the Romance Languages in the Early Middle Ages*, ed. Roger Wright (London: Routledge, 1991), p. 47.

5. *Ibid.*, p. 48.

6. Tore Janson, "Language Change and Metalinguistic Change: Latin to Romance and Other Cases," in Wright, *Latin and the Romance Languages in the Early Middle Ages*, pp. 21–22.

7. Roger Wright, "The Conceptual Distinction between Latin and Romance: Invention or Evolution?" in Wright, *Latin and the Romance Languages in the Early Middle Ages*, p. 109.

8. Peter Burke, "The Uses of Literacy in Early Modern Italy," in *Social History of Language*, eds. Peter Burke and Roy Porter (Cambridge, UK: Cambridge University Press, 1987), pp. 22–23.

9. Wright, "The Conceptual Distinction between Latin and Romance," pp. 104–105.

10. Gottlob Frege, "On Sense and Meaning," in *Translations from the Philosophical Writings of Gottlob Frege*, eds. Peter Geach and Max Black (Totowa, NJ: Rowman and Littlefield, 1980), p. 60. On Frege's theory, see Christian Thiel, *Sense and Reference in Frege's Logic* (Dordrecht, Holland: D. Reidel, 1968), ch. 5.

On its connection with the causal theory of reference, see Nathan U. Salmon, *Reference and Essence* (Princeton, NJ: Princeton University Press, 1981), pp. 11–13 and 31–32. "According to the theory of singular direct reference, proper names and indexical singular terms are either nondescriptional or descriptional in terms of the haecceity of the terms' denotation, the property of *being this very individual*" (*ibid.*, p. 39). This is one of the many points of connection between the new analytical philosophers and Gilles Deleuze and Félix Guattari, who also propose a theory of meaning in terms of haecceities (the "thisness" of an individual) and proper names. See Gilles Deleuze and Félix Guattari, *A Thousand Plateaus* (Minneapolis: University of Minnesota Press, 1987), pp. 262–63.

11. Saul A. Kripke, *Naming and Necessity* (Cambridge, MA: Harvard University Press, 1980), pp. 97–98. Here Kripke develops not only his contribution to the theory of direct reference, but also a separate argument that implies a certain form of "essentialism." That the theory of direct reference does not have to come bundled with a belief in essences is shown by Salmon, *Reference and Essence*, ch. 5.

12. Hilary Putnam, "The Meaning of 'Meaning'," in *Mind, Language and Reality: Philosophical Papers*, 2 vols. (Cambridge, UK: Cambridge University Press, 1980), vol. 2, pp. 225–27. While both the "Twin Earth" argument and the sociolinguistic hypothesis about division of labor and linguistic obligations take care of the notion of "denotation," one may wonder if the concept of "connotation" (so dear to semioticians since Roland Barthes) can also be explained in terms of material labels. In the late 1960s Nelson Goodman developed just such a theory (his theory of "exemplification" as a form of "reverse reference"). See Nelson Goodman, *Languages of Art: An Approach to a Theory of Symbols* (New York: Bobbs Merrill, 1968), ch. 2.

13. On the role of nondiscursive practices in fixing reference, see Ian Hacking, *Representing and Intervening* (Cambridge, UK: Cambridge University Press, 1992), ch. 6. On the relation of causal reference and linguistic history, see Paul M. Lloyd, "On the Names of Languages (and Other Things)," in Wright, *Latin and the Romance Languages in the Early Middle Ages*, pp. 10–11.

14. Lesley Milroy, *Language and Social Networks* (Oxford, UK: Basil Blackwell, 1980), pp. 47–50.

15. *Ibid.*, pp. 21 and 51–52. Social networks capable of acting as enforcement mechanisms must have the additional property of "multiplexity"; that is, the members of the network interact with each other in multiple capacities (kin, workmates, neighbors, partners). This means that their livelihood depends on one another more than if they interacted more impersonally.

16. *Ibid.*, p. 179.

17. Labov, "The Social Setting of Linguistic Change," p. 277; and Samuels, *Linguistic Evolution*, p. 89.

18. Milroy, *Language and Social Networks*, p. 46.

19. John Nist, *A Structural History of English* (New York: St. Martin's, 1976), p. 89.

20. *Ibid.*, p. 91.

21. *Ibid.*, p. 100–101.

22. *Ibid.*, p. 148. Interestingly, the switch from synthetic Old English to analytic Middle English was partly effected through components of language that are usually ignored by formal theories as unimportant: stress and intonation. The English peasants' habit of stressing the first syllables (as in "love," "lover," "loveliness," of Germanic origin, as opposed to "family," "familiar," "familiarity," which are borrowed from Latin) was a powerful selection force in the progressive loss of the syllables at the end of words, which in many cases were inflexions. See *ibid.*, pp. 149–50.

23. Ian F. Hancock, "Recovering Pidgingenesis: Approaches and Problems," in *Pidgin and Creole Linguistics*, ed. Albert Valdman (Bloomington: Indiana University Press, 1977), p. 283.

24. Keith Whinnom, "Lingua Franca: Historical Problems," in Valdman, *Pidgin and Creole Linguistics*, pp. 297–99.

25. Dell Hymes, Preface to *Pidginization and Creolization of Languages*, ed. Dell Hymes (London: Cambridge University Press, 1971), p. 3. There Hymes remarks: "These languages have been considered, not creative adaptations, but degenerations; not systems in their own right, but deviations from other systems. Their origins have been explained, not by historical and social forces, but by inherent ignorance, indolence and superiority."

26. David Decamp, "Introduction: The Study of Pidgin and Creole Languages," in Hymes, *Pidginization and Creolization of Languages*, pp. 19–20.

27. Dell Hymes, "Introduction to Chapter 3," in Hymes, *Pidginization and Creolization of Languages*, p. 79.

28. *Ibid.*, p. 78.

29. *Ibid.*, p. 79.

30. Deleuze and Guattari, *A Thousand Plateaus*, p. 102.

31. Harris, "The Romance Languages," pp. 13–14. On the early rise of Francien and its rivals, see also Ian Parker, "The Rise of the Vernaculars in Early Modern Europe: An Essay in the Political Economy of Language," in *The Sociogenesis of Language and Human Conduct*, ed. Bruce Bain (New York: Plenum, 1983), pp. 342–43, and David C. Gordon, *The French Language and National Identity: 1930–1975* (The Hague, Netherlands: Mouton, 1978), pp. 22–23.

32. Harris, "The Romance Languages," pp. 6–7.

33. Parker, "The Rise of the Vernaculars in Early Modern Europe," p. 344.

34. Harris, "The Romance Languages," p. 18.

35. Parker, "The Rise of the Vernaculars in Early Modern Europe," pp. 337–38.

36. William H. McNeill, *Plagues and Peoples* (Garden City, NJ: Anchor/Doubleday, 1976), p. 162.

37. Samuels, *Linguistic Evolution*, pp. 94–95.

38. Harris, "The Romance Languages," p. 14.

39. *Ibid.*, p. 16; and Gordon, *The French Language and National Identity*, p. 24.

40. Nist, *A Structural History of English*, p. 171.

41. J. L. Austin, *How to Do Things with Words* (Cambridge, MA: Harvard University Press, 1975), p. 26.

42. Deleuze and Guattari, *A Thousand Plateaus*, pp. 80–81. They do not use the word "phase transition" but rather "incorporeal transformation," which amounts to the same thing.

43. Nist, *A Structural History of English*, p. 162.

44. Gordon, *The French Language and National Identity*, p. 23.

45. Nist, *A Structural History of English*, p. 165.

46. Labov, "The Study of Language in Its Social Context," in Labov, *Sociolinguistic Patterns*, pp. 207–12. Sociolinguists attempting to study the casual register using tape recorders constantly run into the following dilemma: the very presence of a microphone tends to provoke the speaker under study to use the formal register. Hence, Labov suggests that to break the constraints of the interview situation one needs to divert attention from speech (by recording samples in situations where emotions run high, and hence where self-monitoring of speech is obstructed) to let the casual register emerge.

47. Jonathan Steinberg, "The Historian and the *Questione Della Lingua*," in Burke and Porter, *The Social History of Language*, p. 204.

48. Ivan Illich, "Vernacular Values and Education," in Bain, *The Sociogenesis of Language and Human Conduct*, p. 467.

49. Einar Haugen, "Dialect, Language, Nation," in *Sociolinguistics*, eds. J. B. Pride and Janet Holmes (Middlesex, UK: Penguin, 1972), pp. 107–108.

50. Illich, "Vernacular Values and Education," p. 470.

51. Einar Haugen, "National and International Languages," in *The Ecology of Language: Collected Papers*, ed. Anwar S. Dil (Stanford, CA: Stanford University Press, 1972), p. 260.

52. Illich, "Vernacular Values and Education," p. 471.

53. Parker, "The Rise of the Vernaculars in Early Modern Europe," pp. 341–42.

54. Bishop Bossuet, quoted in Gordon, *The French Language and National Identity*, p. 26.

55. Antoine Meillet, quoted in Haugen, "National and International Languages," p. 260.

56. Parker, "The Rise of the Vernaculars in Early Modern Europe," pp. 347–48.

57. Nist, *A Structural History of English*, p. 213.

58. *Ibid.*, p. 214.

59. Bill Bryson, *The Mother Tongue* (New York: William Morrow, 1990), p. 93.

60. Samuels, *Linguistic Evolution*, p. 31.

61. *Ibid.*, p. 144. Samuels's account of the evolution of linguistic norms may be easily fit into the model I introduced in the previous chapter; that is, language may be considered to embody a probe head or searching device. However, we saw that as important as the probe head was, an analysis of the space that it blindly explores was also crucial. Yet, this involved bringing into consideration (in the case of organic evolution) material and energetic questions that would not seem to have a counterpart in the world of languages. In other words, I argued that the space that the probe head explores is prestructured by attractors forming so many dynamically stable states (steady states, cyclical states, chaotic states). So the key question here would be, What corresponds to attractors in linguistics? On one hand (in the case of phonemes), one may argue that the nonlinear dynamics of the articulatory system itself is the actual physical and energetic substratum that embodies the attractors. Here, the principle of "least effort" would correspond to the old thermodynamic equilibriums, which are unique and optimal; but further studies of the mouth-teeth-tongue-larynx system may be needed to analyze its operation far from equilibrium. See, for example, H. Herzel, I. Steinecke, W. Mende, and K. Wermke, "Chaos and Bifurcations during Voiced Speech," in *Complexity, Chaos, and Biological Evolution*, eds. Erik Mosekilde and Lis Mosekilde (New York: Plenum, 1991). At the semantic and syntactic levels, the question may be approached in terms of the attractors guiding the computations and other processes in the brain, which is how the connectionist school of Artificial Intelligence frames the question. (Connectionists often talk of "semantic spaces" structured by attractors.)

62. Labov, "The Study of Language in Its Social Context," pp. 221–23.

63. Samuels, *Linguistic Evolution*, p. 173. This quote contains the expression

"free variation," which for linguistic replicators carries the same meaning that "genetic drift" does for genes, that is, random variation. This, however, goes against the treatment of variation by Labov in the note above, for whom certain forms of linguistic variation (e.g., variable rules) are neither random nor due to accidents of contact situations, but *inherent* and *systematic* (what Deleuze and Guattari call "immanent"). See Deleuze and Guattari, *A Thousand Plateaus*, p. 93.

On the general subject of the recruitment of lexical material to play grammatical functions, see also the desemantization and grammaticalization of "to get," in Samuels, *Linguistic Evolution*, p. 58.

64. Labov, "The Study of Language in Its Social Context," pp. 217–18.

65. Deleuze and Guattari, *A Thousand Plateaus*, p. 103.

66. Labov, "The Social Setting of Linguistic Change," p. 298.

67. *Ibid.*, p. 299.

68. Paul M. Hohenberg and Lynn Hollen Lees, *The Making of Urban Europe, 1000–1950* (Cambridge, MA: Harvard University Press, 1985), p. 265.

69. Richard Y. Kain, *Automata Theory: Machines and Languages* (New York: McGraw-Hill, 1972), pp. 4–14.

70. Deleuze and Guattari, *A Thousand Plateaus*, p. 7. The authors' denial of the existence of "linguistic universals" needs to be taken with a grain of salt. Surely the existence (or nonexistence) of universals in language is an empirical question not to be settled by philosophical fiat. Philosophical analysis here is necessary, of course, so that universals as empirically found are not transformed into transcendental entities. The fact that many universals are "statistical universals," that is, common traits shared by languages with above-chance frequency, should already indicate that all we have here is common attractors for a dynamics and not platonic essences. See Joseph H. Greenberg, "Some Universals of Grammar with Particular Reference to the Order of Meaningful Elements," in *Universals of Language*, ed. Joseph H. Greenberg (Cambridge, MA: MIT Press, 1966). Many of the universals mentioned by Greenberg are of the statistical type. Others are of the "implicational" type (that is, if a language has feature *x* then it must also have feature *y*). This latter type of universal may be explained along Zellig Harris's line of thought, according to which linguistic structure grows by accretion: new forms arise by analogy with old ones. More generally, universals of structure may derive from universals of behavior (e.g., common strategies for the exploitation of redundancy in communication). This is particularly clear in the case of pidgins and creoles and their convergence toward universals. See Elizabeth Closs Traugott, "Pidginization, Creolization, and Language," in Valdman, *Pidgin and Creole Linguistics*, p. 82; and Robert Le Page, "Processes of Pidginization and Creolization," in *ibid.*, pp. 229 and 233–34. See also note 108 below.

71. George K. Zipf, *The Psycho-Biology of Language: An Introduction to Dynamic Philology* (Cambridge, MA: MIT Press, 1965), p. 247.

72. Zellig Harris, *A Theory of Language and Information: A Mathematical Approach* (Oxford, UK: Clarendon, 1981), p. 363. Harris's approach connects directly with the other theoretical approaches that I have used here. Deleuze and Guattari, for example, also define linguistic strata in terms of *frequency* as a form of redundancy. See Deleuze and Guattari, *A Thousand Plateaus*, p. 79. And the formalization of Labov's variable rules is also done in terms of frequencies of occurrence. See Labov, "The Study of Language in Its Social Context," p. 231.

73. Harris, *A Theory of Language and Information*, p. 402.

74. *Ibid.*, pp. 329–32.

75. *Ibid.*, pp. 332–34.

76. *Ibid.*, p. 339.

77. Hilary Putnam, "Some Issues in the Theory of Grammar," in *Mind, Language and Reality*, p. 98.

78. Harris, *A Theory of Language and Information*, p. 346.

79. *Ibid.*, pp. 392–94.

80. *Ibid.*, p. 372.

81. *Ibid.*, p. 307.

82. *Ibid.*, p. 309.

83. Mary Douglas, "Introduction to Group/Grid Analysis," in *The Sociology of Perception*, ed. Mary Douglas (London: Routledge and Kegan Paul, 1982), p. 5.

84. *Ibid.*, p. 6.

85. Michael Thompson, "A Three-Dimensional Model," in Douglas, *The Sociology of Perception*, p. 35.

86. David Ostrander, "One- and Two-Dimensional Models of the Distribution of Beliefs," in Douglas, *The Sociology of Perception*, p. 15.

87. Although Mary Douglas's theory is therefore useful in studying the collective character of language's abstract machine, it is important not to see her scheme as providing ammunition for the cultural relativism that I criticized in the previous chapter. While it is true that different intensities of the group and grid parameters yield differing worldviews, this should be seen as the genesis of "moral perception" and not (as Douglas sometimes suggests) of sensory-motor perception. Cultural anthropologists are not the only ones to blame for stressing representational knowledge (worldviews) and ignoring embodied know-how (e.g., skills); some linguists are guilty of this, too.

While it is clear that the availability of linguistic labels does affect somewhat how people relate to the world (for example, by making it *easier* to remember and apply certain categories, i.e., by acting as catalysts), this is a far cry from the claim that we "cut out" the world of perception along purely linguistic lines, as asserted in the Sapir/Whorf hypothesis. In short, it is not the case that Eskimos perceive sixty (or whatever) different types of snow *because* they have sixty different words for snow. Rather, given the key role that snow plays in their nondiscursive daily practices, many synonyms for it can be expected to accumulate and then partially diverge, acquiring subtle shades of meaning. Thus, they have so many words for snow *because* they discriminate many different physically stable states for snow, using embodied intelligence. Besides, I have attempted to show in this book that the world itself is subject to processes of individuation which do not depend on human beings. In other words, reality does not have to wait for humans to sort it out into categories. Sorting processes that produce more or less homogeneous classes of individuals (rocks, species) occur independently of language.

88. Joshua Fishman, "The Impact of Nationalism in Language Planning," in *Language and Society: Collected Papers*, ed. Anwar S. Dil (Stanford, CA: Stanford University Press, 1972), pp. 224–27.

89. Michel Foucault, *Discipline and Punish: The Birth of the Prison* (New York: Vintage, 1979), p. 169.

90. Steven Ross, *From Flintlock to Rifle: Infantry Tactics, 1740–1866* (Cranbury, NJ: Associated University Presses, 1979), pp. 35–39.

91. Gordon, *The French Language and National Identity*, p. 30.

92. Steven Blakemore, *Burke and the Fall of Language: The French Revolution as a Linguistic Event* (Hanover, NH: University Press of New England, 1988), pp. 83–84.

93. *Ibid.*, p. 86.

94. Peter Paret, "Napoleon and the Revolution in War," in *Makers of Modern Strategy: From Machiavelli to the Nuclear Age*, ed. Peter Paret (Princeton, NJ: Princeton University Press, 1986), p. 124.

95. Excerpt from the text of the *levée en mass* of 1793, quoted in William H. McNeill, *The Pursuit of Power: Technology, Armed Force, and Society since A.D. 1000* (Chicago: University of Chicago Press, 1982), p. 192.

96. *Ibid.*, pp. 194–97.

97. Gordon, *The French Language and National Identity*, pp. 30–31.

98. Foucault, *Discipline and Punish*, p. 166.

99. Fernand Braudel, *Capitalism and Material Life, 1400–1800* (New York: Harper and Row, 1973), p. 414.

100. Nist, *A Structural History of English*, pp. 272–75.

101. *Ibid.*, p. 278.

102. *Ibid.*, pp. 280–81.

103. *Ibid.*, p. 305.

104. Hymes, Preface to Hymes, *Pidginization and Creolization of Languages*, p. 5.

105. Decamp, "Introduction: The Study of Pidgin and Creole Languages," p. 19.

106. William Samarin, "Salient and Substantive Pidginization," in Hymes, *Pidginization and Creolization of Languages*, pp. 124–27.

107. Hymes, "Introduction to Chapter 3," in Hymes, *Pidginization and Creolization of Languages*, pp. 67–73; and Keith Whinnom, "Linguistic Hybridization and the Special Case of Pidgins and Creoles," in *ibid.*, p. 104.

108. Derek Bickerton, "Pidginization and Creolization: Language Acquisition and Language Universals," in Valdman, *Pidgin and Creole Linguistics*, pp. 63–64.

109. Traugott, "Pidginization, Creolization and Language," p. 87; and Le Page, "Processes of Pidginization and Creolization," pp. 237–43. One of the puzzles of pidgin linguistics is the close (typological) similarities of all the different pidgins (and creoles) around the world. The convergence toward similar structures is particularly intriguing given the variety of major and minor languages that were involved in their different geneses. An early theory proposed to solve this mystery is the so-called monogenesis hypothesis, according to which all pidgins were really derived from a single one: Sabir. As the European global conquest began, the latest installment of Sabir had a large Portuguese component (reflecting the expanded presence of Lisbon in oceanic navigation), and it was this version that it is supposed to have been brought to West Africa by Portuguese slave traders, and then by the slaves themselves to plantations all over the world. Sabir (or another Portuguese trade pidgin) did indeed replaced Arabic and Malay as the trade language of the Far East during the sixteenth century, and some Spanish creoles (and even Chinese pidgin English) of that region have been shown to derive from that ur-pidgin. Yet many other pidgins (Pitcairnese, Amerindian pidgin English) clearly do not derive from Sabir, and so at least some of these linguistic crystallizations must have occurred independently. Besides, one may wonder why plantation owners would bother purchasing slaves from different linguistic regions in Africa, if the slaves could already communicate with one another in Sabir.

Another explanation for the worldwide convergence involves postulating either linguistic universals (so that the simplification process would become attracted toward some "fixed points" in language space) or behavioral universals, such as common strategies to exploit redundancy or similar ways to deal with situations in which one of the parties does not speak the language well (babies, deaf people, foreigners). In this case it would not be a dynamical situation attracted to fixed points but an oscillatory dynamic (echo-response pattern) in which two or more parties in a heightened state of attention interact using redundancy-exploitation strategies in a constant search of serviceable common elements. Besides this, we must add the institutional context of the slave plantation itself, in which very similar sociolinguistic contact situations were generated.

110. Mervyn C. Alleyne, "The Cultural Matrix of Creolization," in Hymes, *Pidginization and Creolization of Languages*, pp. 182–83.

111. Decamp, "Introduction: The Study of Pidgin and Creole Languages," p. 17.

112. Sidney W. Mintz, "The Socio-Historical Background to Pidginization and Creolization," in Hymes, *Pidginization and Creolization of Languages*, p. 481.

113. *Ibid.*, p. 487.

114. Ali A. Mazrui, *The Political Sociology of English: An African Perspective* (The Hague, Netherlands: Mouton, 1975), pp. 57–58.

115. *Ibid.*, pp. 59–63. In this regard, a distinction can be made between "communalist" and "ecumenical" languages. The former are bound by tribal, regional, or national culture and are therefore *highly absorptive*: speaking those languages as a mother tongue incorporates the speakers into a given culture. Such is the case of Arabic, for instance, which makes anyone who speaks it as a first language an "Arab," regardless of race or color. On the other hand, speaking English does not by itself transform a person into a British citizen, which makes this language less "race bound" (regardless of the fact that the British are more racially exclusive than the Arabs), or, in other words, more ecumenical. French occupies an intermediate position between English and Arabic. Communalist languages tend to foster closer linkages among the different countries that use them: since independence, African colonies that were Francophone have tended to maintain closer ties among one another than have Anglophone colonies. (English-speaking colonies, however, seem to have produced more nationalists than the French-speaking ones, who developed closer emotional ties with the European motherland.) Ecumenical languages, on the other hand, precisely because their adoption carries few national or racial connotations (and hence faces fewer obstacles from local loyalties), tend to spread faster among foreign speakers than communalist languages do (*ibid.* pp. 70–74).

116. *Ibid.*, p. 58.

117. *Ibid.*, p. 136.

118. Wilfred H. Whiteley, "Some Factors Influencing Language Policies in Eastern Africa," in *Can Language Be Planned: Sociolinguistic Theory and Practice for the Developing Nations*, eds. Jaan Rubin and Bjorn H. Jernudd (Honolulu: University of Hawaii Press, 1971), pp. 142–55.

119. Mazrui, *The Political Sociology of English*, pp. 13–14; and Gordon, *The French Language and National Identity*, p. 89. Contemporary African writers who are engaged in deanglicizing English (if not Africanizing it, at least deracializing it, e.g., ridding it of the racist connotations attached to the word "black") answer the former colonialists who decry what is happening to their language that it is not *their* language anymore: the very fact that it has acquired universal currency means it has become everybody's property. French, too, has been appropriated, by African Francophones, particularly those belonging to the Negritude movement (a nationalist literary movement, as opposed to Pan-Negroism, an English-based political movement). African writers involved in this movement attacked French claims to some proprietary right over French, and, as one author puts it, these writers transformed French by "kneading, torturing, disarticulating" it, thereby giving it a new rhythm and a new density. Again, as Deleuze and Guattari said, the more a language becomes major (e.g., English and French as global standards), the more it is appropriated and worked over by various populations who transform segments of it into a minor language. (That is, they "defreeze" its replicating norms and set

them in variation again, to provide the raw materials out of which future languages may evolve.)

120. Nist, *A Structural History of English*, p. 336.

121. Keith Whinnom, "Linguistic Hybridization and the Special Case of Pidgins and Creoles," pp. 92–97.

122. Nist, *A Structural History of English*, pp. 347–50 and 366–67. These regionalisms

> now constitute three major speech areas in the United States: Northern, Midland, and Southern. These speech areas foster their own regional subdivisions, which at times have little to do with geographical location. Thus the pronunciation habits of the Southwest area of Arizona, Nevada, and California are generally of Northern derivation, whereas the speech patterns of the Northwest (Montana, Idaho, Oregon, and Washington) are basically of Midland origin. West Texas shows the dominance of Appalachian; East Texas speech is an outgrowth of Southern. Since Appalachian is a regional version of Midland, the differences between the pronunciations of East and West Texas are marked. (pp. 366–67)

Thus, even though each of the main regional variants had a definite geographical center (Boston for Northern, New York City for Midland, and cities like Richmond, Atlanta, and New Orleans for Southern), the role of railroads in the settling of the central and western areas of the country, as well as their contribution to postsettlement migratory flows and to the general mobility of the population, meant that geography alone was not to determine the distribution and accumulation of linguistic variants in the United States.

123. Ken Ward, *Mass Communications and the Modern World* (London: Macmillan Education, 1989), p. 36.

124. *Ibid.*, p. 23.

125. *Ibid.*, pp. 91, 97–98, and 121–23.

126. *Ibid.*, p. 33.

127. Jonathan Fenby, *The International News Agencies* (New York: Schocken, 1986), pp. 24–25 and 33–37.

128. *Ibid.*, pp. 62–63.

129. *Ibid.*, p. 88.

130. Nist, *A Structural History of English*, pp. 306–307.

131. *Ibid.*, p. 383.

132. Tony Crowley, *Standard English and the Politics of Language* (Champaign: University of Illinois Press, 1989), pp. 215–17.

133. *Ibid.*, p. 252.

134. *Ibid.*, p. 159.

135. Samuels, *Linguistic Evolution*, p. 108.

136. Bjorn H. Jernudd, "Notes on Economic Analysis for Solving Language Problems," in Rubin and Jernudd, *Can Language Be Planned*, pp. 272–73.

137. Fishman, "National Languages and Languages of Wider Communication in the Developing Nations," in Dil, *Language and Society*, p. 197.

138. Fishman, "The Impact of Nationalism in Language Planning," pp. 228–30.

139. Fishman, "National Languages and Languages of Wider Communication in the Developing Nations," pp. 192–93 and 222.

140. The case of Turkey is particularly interesting because, in its pre–World War I incarnation (the Ottoman Empire), Turkey was by no means a periphery to Europe, but a participant on equal terms on the international scene (like Japan later on, but unlike India or China). Turkey had undergone a first "civilization rite of passage" in medieval times as it became an Islamic country. Turkish had become the language of the peasantry, while the urban elites spoke Osmanlica, an amalgam of Arabic, Persian, and Turkish. In the nineteenth century, the Ottomans became aware of a growing vocabulary gap between Osmanlica and the European languages, particularly in the military and industrial technical registers. But the Arabic elements in their linguistic mixture made translation and adaptation of the new words hard, and pressures for the de-Arabization and de-Persiafication of their language began to be felt. The intensification of nationalist feelings after the 1908 revolution and the shock of World War I accelerated the process, and Turkey underwent a second rite of passage, this time to cut off all ties to its Islamic past and to completely secularize (and standardize) a revived Turkish language. Although Turkey did possess a grand tradition to legitimate the process, the presence of Islamic elements in that tradition meant that the elites would have to manipulate history to justify the radical changes their national language would have to undergo, such as romanization of its script and enrichment of its lexical reservoir through borrowing. A linguistic theory was concocted (the "sun language" theory), according to which Turkish was the mother of all languages, and hence the borrowing of foreign words could be justified on the grounds that those lexical items had once belonged to ancient Turkish. See Charles F. Gallagher, "Language Reform and Social Mobilization in Turkey," in Rubin and Jernudd, *Can Language Be Planned*, pp. 161–66.

141. Those countries lacking a unifying tradition had to face a more difficult set of choices. In particular, in selecting a candidate for standardization, they could either pick the language of one of their elites or what is called a "language of wider circulation," which can be the language of the ex-colonial masters (English, French, Spanish, or Dutch) or a local lingua franca (Swahili, Malay). The first choice meant favoring the members of a particular prestigious group at the expense of other, perhaps equally prestigious ones, and so it immediately confronted opposition from the excluded elites. Picking a colonial language contradicted some of the goals of nationalism (yet this choice was made by quite a few countries), so whenever a lingua franca was available it became a serious contender for a national standard.

Indonesia offers a good example of this strategy. This archipelago provides many geographical barriers to the spread of linguistic replicators, and thus, by favoring isolation over contact, it gave rise to over two hundred separate languages. In these conditions a lingua franca (Malay) emerged early on for trade and political interaction. Much as the British and Germans picked Swahili as one of their colonial administration languages, so the Dutch here selected Malay, further enhancing its popularity. Although efforts at standardization began early in the 1930s, it was the Japanese who institutionalized the project when they occupied the islands during World War II, banishing Dutch, establishing a committee with the aim of creating a grammar and a dictionary, and making Malay a medium of instruction. Hence, in this case, instrumentality and "rational" planning (routinization) outweighed authenticity as a selection criterion for the standard, since a less prestigious variant was picked over an elite variety on the grounds of its efficiency and currency in communication. See S. Tadkir Alissahbana, "Some Planning Processes in the Development of the Indonesian-Malay Language," in Rubin and Jernudd, *Can Language Be Planned*, pp. 180–84.

142.   While Turkey and Indonesia, in their different ways, arrived at a single national standard, other countries faced with several rival traditions were forced to make compromises. Ethiopia, for example, today has five major languages, including Amharic (an indigenous standard, with a writing system and literature dating back to the fourteenth century) and English (the language of instruction and international communication). It also has special liturgical languages (Arabic and Geez, each a sacred language for a different grand tradition) that enjoy as much prestige as the major ones. See C. A. Ferguson, "The Role of Arabic in Ethiopia: A Sociolinguistic Perspective," in Pride and Holmes, *Sociolinguistics*, p. 114.

India, on its side, also has two competing grand traditions (Islam and Hinduism) and sixteen different languages cutting across religious boundaries: Hindi and Urdu, for example, belong to the same linguistic family, but the latter is more Islamized, making use of many Persian words, while the former retains its ties to Sanskrit. To this day their rivalry continues, and India has been forced to recognize several standards. See Jyotirindra Das Gupta, "Religion, Language and Political Mobilization," in Rubin and Jernudd, *Can Language Be Planned*, pp. 55–60.

143.   The definition of "standardization" as composed of "codification" and "elaboration" appears in Haugen, "Dialect, Language, Nation," pp. 107–108.

144.   Gordon, *The French Language and National Identity*, p. 42.

145.   *Ibid.*, p. 45.

146.   *Ibid.*, p. 48.

147.   *Ibid.*, p. 56.

148.   *Ibid.*, pp. 97–98.

149.   *Ibid.*, p. 42.

150.   Robert Phillipson, *Linguistic Imperialism* (Oxford, UK: Oxford University Press, 1993), p. 111.

151. *Ibid.*, p. 28 (on English schooling) and pp. 113–14 (on French).

152. *Ibid.*, pp. 300–302.

153. Fenby, *The International News Agencies*, p. 3.

154. *Ibid.*, p. 159.

155. *Ibid.*, p. 117.

156. Howard Rheingold, *The Virtual Community: Homesteading on the Electronic Frontier* (New York: Harper Perennial, 1994), p. 130.

157. *Ibid.*, ch. 8.

158. Mazrui, *The Political Sociology of English*, p. 75.

159. Bryson, *The Mother Tongue*, p. 184.

This inclination to hack away at English words until they become something like native products is not restricted to the Japanese. In Singapore transvestites are known as *shims*, a contraction of *she-hims*. Italians don't go to a nightclub, but just to a *night* (often spelled *nihgt*), while in France a self-service restaurant is simply *le self*. European languages also show a curious tendency to take English participles and give them entirely new meanings, so that the French don't go running or jogging, they go footing.... The Germans are particularly inventive at taking things a step further than it ever occurred to anyone in English. A young person in Germany goes from being in his teens to being in his *tweens*, a book that doesn't quite become a best-seller is instead *ein steadyseller*, and a person who is more relaxed than another is *relaxter*.

160. Rheingold, *The Virtual Community*, p. 234. The French and Japanese, having experimented with their own national computer networks, may not welcome the potential heterogenizing effects of opening up to the international meshwork. Rheingold observes:

> The challenge now confronting France, after more than a decade of this experiment [i.e., Minitel], has some of the challenges now facing Japan. Because of Japanese restrictions on their own communications market, they were late to develop; now they are faced with the growth of Internet and the cultural conflicts that full Internet access would precipitate. France closely guards against cultural intrusion, as in its *dirigiste* attempts to control the French language through the Academie. Fear of American competition and distrust of the Internet experiment colored the decisions that went into the original Teletel design. The tiny screens and almost unworkable keyboards of the millions of Minitels now in use are clearly inadequate in the age of high-bandwidth communications and powerful desktop computers. Will France redesign its user interface, and thus leap forward again, or will it be chained to the investment in crude terminals that was revolutionary ten years ago? And if France leaps ahead ... will that French network wall itself off from the Net, the way it has done in the past?

Or will it join the Net and give it more of a French flavor — and inevitably, discover that the Net has changed French culture, in ways that are not all pleasant? (p. 234)

161. *Ibid.*, p. 88.

162. Deleuze and Guattari, *A Thousand Plateaus*, p. 500. The original sentence contains the words "smooth space" not "meshwork," but, arguably, both refer to basically the same thing. Also see note 2 below.

CONCLUSION AND SPECULATIONS

1. Gilles Deleuze and Félix Guattari, *A Thousand Plateaus* (Minneapolis: University of Minnesota Press, 1987), p. 159.

2. While the term "Body without Organs" was first used in a philosophical context by Gilles Deleuze (borrowing from Artaud), the almost synonymous "machinic phylum" seems to have been coined and first used by Guattari, in Félix Guattari, "The Plane of Consistency," in *Molecular Revolution* (New York: Penguin, 1984), p. 120. I do not claim that the two terms are strictly synonymous (although *I* use them that way). Rather, these philosophers, instead of building *one theory*, are attempting to create a *meshwork of theories*, that is, a set of partially overlapping theories. Hence, nearly synonymous key concepts (BwO, phylum, smooth space, rhizome) do not exactly coincide in meaning but are slightly displaced from one another to create this overlapping effect. The point remains that the referents of these labels, not the labels themselves, are what matters.

3. Deleuze and Guattari, *A Thousand Plateaus*, p. 164.

4. On the consequences for human history of the breakup of Pangaea and the consequent distribution of domesticable species, see Jared Diamond, *The Third Chimpanzee* (New York: HarperCollins, 1982), pp. 237–39. Diamond enumerates the preconditions for domesticability which apply to most domesticates, except cats (living in herds, low-intensity territorial behavior and flight reflexes, etc.), and discusses the effects of this biogeographical accident on the colonial confrontations between Eurasia and the rest of the world. For instance, of the horse he writes: "The military value of the horse is specially interesting in illustrating what seemingly slight differences make one species uniquely prized and another useless. . . . Of the seventeen living species [belonging to the same order as the horse] all four tapirs and all five rhinos, plus eight of the eight wild horse species, have never been domesticated. Africans or Indians mounted on rhinos or tapirs would have trampled any European invaders, but it never happened" (*ibid.*, p. 239).

5. Deleuze and Guattari, *A Thousand Plateaus*, pp. 6–7. Here Deleuze and Guattari write: "Rats [in their pack form] are rhizomes. Burrows are too, in all their functions of shelter, supply, movement, evasion, and breakout. . . . When rats swarm over each other."

6. *Ibid.*, p. 153.

7. *Ibid.*, pp. 220–21. Here Deleuze and Guattari write:

> [W]e must introduce a distinction between the two notions of *connection* and *conjugation* of flows. "Connection" indicates the way in which decoded and deterritorialized flows boost one another, accelerate their shared escape.... [T]he "conjugation" of these same flows, on the other hand, indicates their relative stoppage, like a point of accumulation that plugs or seals the lines of flight, performs a general reterritorialization, and brings the flows under the dominance of a single flow capable of overcoding them. But it is precisely the most deterritorialized flow, under the first aspect, that always brings about the accumulation or conjunction of the processes, determines the overcoding, and serves as the basis for a reterritorialization under the second aspect.... For example, the merchant bourgeoisie of the cities conjugated or capitalized a domain of knowledge, a technology, assemblages and circuits into whose dependency the nobility, Church, artisans, and even peasants would enter. It is precisely because the bourgeoisie was a cutting edge of deterritorialization, a veritable particle accelerator, that it also performed an overall reterritorialization.

Despite the fact that their philosophical work represents an intense movement of destratification, Deleuze and Guattari seem to have preserved their own stratum, Marxism, which they hardly touch or criticize (except in the obvious ways—i.e., party-oriented versions of it). They retain the concept of "mode of production" and of "capitalist system" defined in a top-down way as an axiom of decoded flows. It seems to me that it would be useful to push their own line of flight even further, abandoning molar concepts and dealing exclusively with multiplicities, in this case, populations of institutions which do not form an overall system.

8. *Ibid.*, p. 69.

9. Gilles Deleuze, *Foucault* (Minneapolis: University of Minnesota Press, 1986), p. 47. Given that the development of languages (via sorting and consolidation) may embody the same abstract diagram as rocks and biological species, it follows that both linguistic structures and their referents in reality may be *isomorphic*. Moreover, if this double articulation also links institutional organizations (acting as sorting devices for human bodies) and the discourses generated in and by these institutions (as suggested by Deleuze's reading of Foucault), then this isomorphism may be said to exist along several dimensions. This suggests the possibility of a rather novel approach to the analysis of theories of truth. Specifically, sentences would not connect to reality via correspondences but isomorphisms:

> Not only do linguistic variables of expression enter into relations of formal opposition or distinction favorable for the extraction of constants; non-linguistic variables of content do also. As Hjelmslev notes, an expression is divided, for example, into phonic units in the same way a content is divided into social, zoological, or physical units....

The network of binarities, or arborescences, is applicable to both sides [i.e., content and expression]. There is, however, no analytic resemblance, correspondence or conformity between the two planes. But their independence does not preclude isomorphism.... (Deleuze and Guattari, *A Thousand Plateaus*, p. 108)

10. *Ibid.*, p. 70.

11. Gilles Deleuze and Félix Guattari, *Anti-Oedipus* (New York: Viking, 1977), p. 42.

12. Michel Foucault, *Discipline and Punish: The Birth of the Prison* (New York: Vintage, 1979), p. 139.

13. Deleuze and Guattari, *A Thousand Plateaus*, pp. 160–61.

14. On the systematic disregard for anything but linear equations in science, see for example Ian Stewart, *Does God Play Dice? The Mathematics of Chaos* (Oxford, UK: Basil Blackwell, 1989), p. 83. There Stewart writes:

So docile are linear equations that the classical mathematicians were willing to compromise their physics to get them. So the classical theory deals with *shallow* waves, *low*-amplitude vibrations, *small* temperature gradients. So ingrained became the linear habit that by the 1940's and 1950's many scientists and engineers knew little else.... Linearity is a trap. The behavior of linear equations ... is far from typical. But if you decide that only linear equations are worth thinking, self-censorship sets in. Your textbooks fill with the triumphs of linear analysis, its failures buried so deep that the graves go unmarked and the existence of the graves goes unremarked.

Further discussion of this crucial aspect of the sociology of science (supplemented with some anecdotal evidence) may be found in James Gleick, *Chaos* (New York: Viking, 1987), pp. 35–39. A more philosophical discussion and some harder evidence of "repression of the nonlinear" may be found in Stephen H. Kellert, *In The Wake of Chaos* (Chicago: University of Chicago Press, 1993), ch. 5. On the proliferation of equilibrium models in the social sciences, see Cynthia Eagle Russett, *The Concept of Equilibrium in American Social Thought* (New Haven, CT: Yale University Press, 1968).

15. J. E. Gordon, *The Science of Structures and Materials* (New York: Scientific American Books, 1988), p. 200.

16. Deleuze and Guattari, *A Thousand Plateaus*, p. 330.

17. *Ibid.*, p. 336.

18. *Ibid.*, p. 61.

19. Deleuze, *Foucault,* p. 93. There Deleuze writes: "[F]or Foucault as much as Nietzsche, it is in man himself that we must look for the set of forces and functions which resist the death of man. Spinoza said there was no telling what the human body might achieve, once freed from human discipline. To which Foucault replies

that there is no telling what man might achieve 'as a living being,' as the set of forces that resist."

However, in a footnote elsewhere, Deleuze and Guattari disagree with the idea that this destratifying potential may be reduced to acts of political "resistance": "Our only points of disagreement with Foucault are the following[:] the diagram and the abstract machine have lines of flight which are primary, which are not phenomena of resistance or counterattack in an assemblage, but cutting edges of creation and deterritorialization" (Deleuze and Guattari, *A Thousand Plateaus*, p. 531).

This edition designed by Bruce Mau Design Inc.,
Bruce Mau with Chris Rowat
Text typeset in News Gothic by Archetype.